Withdrawn

What Can I Do Now?

Travel and Tourism

Second Edition

Books in the
What Can I Do Now? Series

Animal Careers
Animation
Art
Business and Finance
Computers
Education
Engineering, Second Edition
Environment, Second Edition
Fashion
Film
Health Care
Journalism, Second Edition
Law
Music
Nursing, Second Edition
Radio and Television, Second Edition
Safety and Security, Second Edition
Science
Sports, Second Edition
Travel and Tourism, Second Edition

What Can I Do Now?

Travel and Tourism

Second Edition

Ferguson Publishing
An imprint of Infobase Publishing

What Can I Do Now? Travel and Tourism, Second Edition

Copyright © 2010 by Infobase Publishing

Ferguson
An imprint of Infobase Publishing
132 West 31st Street
New York NY 10001

Library of Congress Cataloging-in-Publication Data

What can I do now? Travel and tourism. —2nd ed.
    p. cm. — (What can I do now? series)
  Includes bibliographical references and index.
  ISBN-13: 978-0-8160-8078-6 (hardcover : alk. paper)
  ISBN-10: 0-8160-8078-X (hardcover : alk. paper) 1. Tourism—Vocational guidance—Juvenile literature. I. Ferguson Publishing. II. Title: Travel and tourism.
  G155.5.W53 2010
  910.23—dc22
                    2009039497

Ferguson books are available at special discounts when purchased in bulk quantities for businesses, associations, institutions, or sales promotions. Please call our Special Sales Department in New York at (212) 967-8800 or (800) 322-8755.

You can find Ferguson on the World Wide Web at http://www.fergpubco.com

Text design by Kerry Casey
Composition by Mary Susan Ryan-Flynn
Cover printed by Sheridan Books, Ann Arbor, MI
Book printed and bound by Sheridan Books, Ann Arbor, MI
Date printed: March 2010
Printed in the United States of America

10 9 8 7 6 5 4 3 2 1

Contents

Introduction

If you are considering a career in travel and tourism—which is presumably the reason you're reading this book—you must realize that the better informed you are from the start, the better your chances are of having a successful, satisfying career.

There is absolutely no reason to wait until you get out of high school to "get serious" about a career. That doesn't mean you have to make a firm, undying commitment right now. Gasp! Indeed, one of the biggest fears most people face at some point (sometimes more than once) is choosing the right career. Frankly, many people don't "choose" at all. They take a job because they need one, and all of a sudden 10 years have gone by and they wonder why they're stuck doing something they hate. Don't be one of those people! You have the opportunity right now—while you're still in high school and still relatively unencumbered with major adult responsibilities—to explore, to experience, to try out a work path. Or several paths if you're one of those overachieving types. Wouldn't you really rather find out sooner than later that you're not cut out to be a pilot after all, that you'd actually prefer to be a travel agent, or a flight attendant, or an adventure-travel specialist?

There are many ways to explore the travel and tourism industry. What we've tried to do in this book is give you an idea of some of your options. The What Do I Need to Know About Travel and Tourism? section will give you an overview of the field—a little history, where it's at today, and promises of the future; as well as a breakdown of its structure—how it's organized—and a glimpse of some of its many career options.

The Careers section includes 10 chapters, each describing in detail a specific career in the travel and hospitality industry: adventure-travel specialist, bed and breakfast owner, cruise ship worker, flight attendant, hotel concierge, hotel manager, pilot, tour guide, travel agent, and travel writer. The educational requirements for these specialties range from high school diploma to bachelor's degree. These chapters rely heavily on firsthand accounts from real people on the job. They'll tell you what skills you need, what personal qualities you must have, and what the ups and downs of the jobs are. You'll also find out about educational requirements—including specific high school and college classes—advancement possibilities, salary ranges, and the future employment outlook.

In keeping with the secondary theme of this book (the primary theme, for those

1

of you who still don't get it, is "You can do something now"), Section 3, Do It Yourself, urges you to take charge and learn about travel-related careers on your own and start your own programs and activities where none exist—school, community, or the nation. Why not?

The real meat of the book is in Section 4, What Can I Do Right Now? This is where you get busy and *do something*. The chapter "Get Involved" will clue you in on the obvious volunteer and intern positions, the not-so-obvious summer camps and summer college study, and other opportunities.

"Read a Book" is an annotated bibliography of books (some new, some old) and periodicals. If you're even remotely considering a career in this field, reading a few books and checking out some magazines or professional journals is the easiest thing you can do. Don't stop with our list. Ask your librarian to point you to more materials. Keep reading!

While we think the best way to explore travel- and tourism-related careers is to jump right in and start doing it, there are plenty of other ways to get into the hospitality mind-set. "Surf the Web" offers a short annotated list of Web sites where you can explore everything from job listings (start getting an idea of what employers are looking for now), to educational requirements and school listings, to on-the-job accounts.

"Ask for Money" is a sampling of scholarships for people who are interested in pursuing travel- or tourism-related careers. You need to be familiar with these because you're going to need money for school. You have to actively pursue scholarships; no one is going to come up to you one day and present you with a check because you're such a wonderful student. Applying for scholarships is work. It takes effort. And it must be done right and often as much as a year in advance of when you need the money.

"Look to the Pros" is the final chapter. It lists professional organizations you can turn to for more information about accredited schools, education requirements, career descriptions, salary information, union membership, job listings, scholarships, and more. Once you become a college student in a travel- or tourism-related field, you'll be able to join many of these; others can be joined by people of any age. Time after time, professionals say that membership and active participation in a professional organization is one of the best ways to network (make valuable contacts) and gain recognition in your field.

High school can be a lot of fun. There are dances and football games; maybe you're in band or play a sport. Great! Maybe you hate school and are just biding your time until you graduate. That's too bad. Whoever you are, take a minute and try to imagine your life five years from now. Ten years from now. Where will you be? What will you be doing? Whether you realize it or not, how you choose to spend your time now—studying, playing, watching TV, working at a fast food restaurant, hanging out, whatever—will have an impact on your future. Take a look at how you're spending your time now and ask yourself, "Where is this getting me?" If you can't come up with an answer, it's probably "nowhere." The choice is yours. No one is going to take you by the hand and lead you in the "right" direction. It's up to you. It's your life. You can do something about it right now!

SECTION 1

What Do I Need to Know About Travel and Tourism?

The idea of travel is a glamorous one. When you think of travel, you may picture yourself sipping café au lait in a Parisian sidewalk café. You may imagine going on safari in Africa. Or perhaps your idea of travel is merely relaxing in a fancy five-star hotel that places foil-wrapped, upscale chocolates on your pillow. Or a more realistic one of traveling on a family vacation and spending the night at a quaint mom-and-pop operation—you know, the kind with a fenced (chain-linked, of course) in-ground pool. Maybe you'd rather relax in a deck chair on a cruise ship or go deep-sea diving in the Caribbean.

Whatever your travel fantasy is, however, it probably doesn't include some important details—like how you're going to get there, where you're going to stay, finding the best price on airfare and car rentals, or how your money converts to the local currency. Even though these details may not be included in your travel fantasy, they are nonetheless essential elements of any trip. The individuals we rely on to handle these details, and to get us comfortably from point A to point B, are those who work in the travel and tourism industry.

Not all careers in travel and tourism require that you actually travel. There is, in fact, a very broad range of career opportunities that require no travel at all. It is perhaps easiest to look at the travel and tourism industry as broken down into two major divisions: planning and execution. That is, there are thousands of people working just to help travelers plan their trips. These individuals check fares, book seats on airplanes, reserve rental cars and hotel rooms, and provide information on planned destinations—but they have nothing to do with the actual trip itself. These are the people working in the planning division.

Workers in the execution division actually deal with travelers while they are on their trips. They are the pilots who fly the airplanes, the flight attendants who seat and serve air travelers, and the tour guides who oversee groups of travelers. They are the concierges in hotels who give tips on hot destinations, the executive housekeepers who provide you with a clean and comfortable room, and the restaurant managers who make sure you have a good meal at the start or end of your exciting vacation or business day.

Many jobs in the execution division require you to be very mobile. Other jobs, however, such as travel agent or airline or rental car reservation agent, allow you the convenience of a routine, one-location job. Depending upon your personal desires and needs, you can pursue a travel and tourism job that turns you into a world traveler or one that lets you come home for dinner every night.

Many careers in the travel branch of the travel and tourism industry rarely require a college degree; instead, they usually require specialized training. In some cases, this training is provided on the job. Most airlines, for example, have extensive training programs for their employees— from baggage handlers to flight attendants to ticket agents. In other cases, training can be obtained through independent schools or training programs. There are

Lingo to Learn

Airlines Reporting Corporation (ARC) An independent corporation created by domestic airlines that governs and regulates payments to airlines and commission fees to travel agencies.

city codes Three-letter codes used to uniquely identify cities and/or their airports; for example, "DCA" is used for Ronald Reagan International Airport in Arlington, Virginia.

customs A government agency that monitors the flow of goods and substances into and from a country.

ground arrangements Services covering the land portion of a trip, such as lodging, transportation, sightseeing, and meals.

hub-and-spoke A system many airlines have adopted to maximize the amount of time their planes spend in the air; designating certain cities as hubs, scheduling many flights to them, and offering connecting flights from the hubs to smaller cities, which can be served by smaller aircraft.

itinerary The route of travel.

package A travel product bundling several distinct elements, such as air travel, a rental car, and a hotel.

passport A document identifying an individual as a citizen of a specific country and attesting to his or her ability to travel freely.

terminal An airport, train station, or bus station.

visa A document or, more frequently, a stamp in a passport authorizing the bearer to visit a country for specific purposes for a specific length of time.

---

many such independent training programs for travel agents. And even though a degree is rarely required, there are certainly many college graduates working in the field as well as in management positions. In travel, as in most industries, the more education you have, the better your chances for advancement.

Though a degree in hotel management is increasingly preferred for upper management, accounting, marketing and sales, and other executive positions, a college education is not a requirement for most other jobs in the hospitality branch of the travel and tourism field. Bellhops, housekeepers, switchboard operators, and desk clerks are among the many entry-level jobs available to recent high school graduates, or even students wishing to hold part-time jobs while in high school. This is an industry that truly rewards experience. It is not uncommon to hear of former busboys or bellhops rising to the rank of banquet manager or general manager. To accomplish this purpose, hotels and motels run an intricate system of keeping house. There is a job for every talent and interest within the growing hotel and motel industry. If you are skillful in organizing and helping people, maybe a job as a desk or reservation clerk is for you. Do you feel most at ease when planning a dinner or party? If so, then a position in the banquet department sounds just right. Are you "in the know" when it comes to the best restaurant or hottest ticket in town? Then you should find out what other interests and skills it takes to become a concierge. Most important, whatever department grabs your attention, remember that the lodging trade is all about serving the guest. Candidates with a pleasant personality and desire to help will succeed. Hotheads need not apply!

Whatever your talent, whatever your level of education, read on and learn if a career in travel and tourism is for you. This book will try to give you a broad overview of the many career options available to you and how you can start preparing for them right now.

GENERAL INFORMATION

Hotels and motels have always existed in some form or another as long as people have needed to take trips that require an overnight stay. These structures were built along trading routes followed by travelers long before the first roads were built. The earliest lodging places, called inns, most likely offered little more than a sheltered area, usually by a source of water. In many parts of the world, it was the custom for people to offer a resting place to weary travelers in their homes.

When the Roman Empire was at the peak of its power in A.D. 100, it built the

first great system of roads. These cobblestone roads were the most extensive, well-constructed roads ever seen, covering more than fifty thousand miles. While this road system was constructed by and for the Roman troops, it was used mainly by people in commerce and trade to transport goods between Rome and other cities. Eventually, inns and restaurants developed along the roads to accommodate the needs of travelers on long journeys. As transportation improved, the number of people traveling increased. Pilgrimages to the holy lands, sacred places, and the sites of miracles became common undertakings for the population. Long voyages of several hundred and several thousand miles were made by those in search of adventure or commerce. Eventually, travel began to be accepted for the sake of entertainment and enjoyment. It provided a diversion for those with sufficient time and money.

In America the first public inn was built in Jamestown, Virginia, in 1607. Most of the early American hotels were established on the East Coast, where travelers from Europe would disembark. Inns, taverns, and ordinaries, as they were called in the southern colonies, appeared along canals, rivers, seaports, and roads. As the country expanded into the western territories, farmhouse inns were maintained along the stagecoach routes.

Technology began to make travel much easier and more affordable in the 19th century. Starting in the late 18th and early 19th centuries, inventors competed to develop the first steam-powered locomotives. By the 1830s, the first workable engines and rail lines carried goods and passengers throughout the Atlantic states. In 1852, the first train reached Chicago, and soon the new towns in the Mississippi Valley had railway service to the Eastern seaports. The West and East were finally linked by a rail line in 1869, when the First Continental Railway was completed. The rise of railroads increased the demand for hotels and inns in both Europe and the United States. Having a railroad stop was usually a boon to small towns. Hotels were usually located close to the train station; they also grew larger to accommodate the increasing numbers of travelers.

Starting in the second half of the 19th century, steamships gradually replaced sailing vessels on the world's trade and passenger routes. Passenger travel between Europe and the States increased. Luxury cruise ships—to carry passengers to and from Europe—were built with the best accommodations that could be put on a ship. These vessels featured orchestras, ballrooms, fine dining halls, and such. The *Queen Elizabeth* was probably one of the best-known and most-traveled luxury liners. Ship travel was intended to be leisurely. Crossing the Atlantic by ship took two weeks.

With the new methods of travel, tourism was no longer limited to the "idle rich." The working class could afford train fare to the countryside or the big cities. For Americans, a trip to Europe became an achievable goal; although the cost was still quite high, it was within the grasp of enough people to save for and plan the trip. In the late 19th and early 20th centuries, a trip abroad was a regular gift to the graduating college student. Referred to commonly as a "grand tour," it was seen as a reward as well as a learning experience for the college graduate.

A new age of travel began when the Wright brothers made the first successful powered flight of a heavier-than-air craft in 1903. Within a few decades, the airplane had secured its place as a vital means of transportation. As the airline industry developed, advancements in plane design allowed for a greater number of passengers on a greater number of routes. Small cities were able to establish airports for smaller vehicles, and large cities found themselves with several runways and substantial air traffic.

As the number of travelers increased, hotel and motel owners began to compete for customers by creating more luxurious and service-oriented establishments. Lodgings began to offer more than a bed, a meal, and a roof over one's head. People began to have parties and meetings at inns. The size of the average hotel increased. The largest hotels would have hundreds of rooms. During the 19th century, many luxury hotels were established to cater to the well-to-do traveler. These hotels featured dining rooms, ballrooms, shops, billiard and sitting rooms, and other amenities, in addition to many clean, well-appointed rooms. Often as large and as opulent as palaces, these *hotels de grand luxe* appeared in the major cities of the United States and Europe. The Ritz hotels of London, New York, and Paris, the Palace Hotel of San Francisco, and the Waldorf Astoria of New York were famous for their luxurious accommodations.

Due to the growing interest in and popularity of commercial travel, the need for people who could plan trips increased. Travel agents who knew which hotels were good, how to get reservations, and how to make travel plans found themselves increasingly in demand. In England, Thomas Cook began his business with the guided tour in 1841. He specialized in excursions that serviced hundreds of thousands of people a year. Exotic places like Egypt and the Orient were open to travelers on the Cook tours. Travel agencies developed everywhere in the West. Travel specialists who could arrange tours and travel guides who knew the ins and outs of faraway places became sought-after businesspeople. The booming travel industry relied on experts to steer tourists to their establishments.

Until the 20th century, travelers had little to choose between the luxury hotels on the one end and the inexpensive hotels, which were not always very clean or comfortable, on the other. This changed when Ellsworth Statler began building his chain of mid-priced Statler Hotels, which set a new standard for the quality, amenities, and service the middle-class traveler could expect. Statler's hotels featured clean, comfortable rooms, each with a private bath, telephone, radio, full-length mirror, and closet, for a modest price. Coordinating linens, china, and silverware were used in each Statler hotel restaurant, as well as standardized recipes.

Freestanding (or self-contained) restaurants were once associated almost exclusively with hotels. Prohibition and the 1920s dramatically altered that association.

Before the Volstead Act, which outlawed the consumption of alcohol in the United States, hotel guests could relax with a drink in the downstairs bar or restaurant. Denied their drinks during Prohibition, they left their hotels in the evening to scout the neighborhood for one of the many speakeasies that served liquor illegally. Most of the speakeasies also provided food, both to cover their illegal activities and to please their customers. Thus, Prohibition helped drive a wedge between the hotel and the restaurant, two institutions that traditionally had coexisted for mutual profit.

More dramatic changes for the hotel profession came with the rise of the automobile. A new type of hotel appeared to cater to the increasing mobility of the public. Motor inns and travel courts, often built by farmers where their land faced onto one of the new roads being built all across the country to meet the needs of the motoring population, offered far simpler accommodations than a hotel. Despite an early reputation as a gathering place for thieves and other undesirable people, motor inns—later called motels (a contraction of the words motor and hotel), soon drew off much of the business from the city-based hotels. Motels gained an image as inexpensive, simple lodging places offering convenient automobile parking for guests. However, modern motels have become larger, fancier establishments with conference and ballroom facilities, as well other amenities comparable to larger hotels, so the gap in quality between the two is much smaller.

The hotels responded to this new competition by building or expanding their facilities to accommodate large meetings and conventions. More and more self-standing restaurants appeared, many offering simple, quick fare to motorists. The automobile made even the most remote locations accessible, and resorts were built to cater to every type of vacationer.

The biggest boom in the hotel and motel industry came with the end of World War II. More and more people were purchasing cars, and with a healthy economy and growing amounts of leisure time, more people were traveling, dining out, and vacationing than ever before. Hotel chains such as the Statler and Hilton hotels were soon joined by chains of motels, such as Howard Johnson and Holiday Inn. Entire cities became devoted to the hotel industry, such as the gambling city of Las Vegas, or seaside resort towns such as Atlantic City. The growth of the airline industry in the 1950s made long-distance travel still more practical and created a demand for new hotels, now located near the airports.

One trend in the hotel and motel industry of the last several decades has been toward consolidation. Today, more than 50 percent of the hotel and motel beds in the United States are controlled by about 25 companies. The world's largest lodging group is InterContinental Hotels Group, with nearly 620,000 guest rooms at its more than 4,150 properties.

A second trend in the hospitality industry is the growing popularity of bed and breakfasts (B & Bs), which offer a personal touch to the tourist or businessperson tired of staying in impersonal chain hotels

> ## Monasteries: Our Early Hotels
> ● ● ● ● ● ● ● ● ●
>
> In addition to our earliest inns, travelers could also find lodging at monasteries. Eventually monasteries developed separate dormitory lodgings for such visitors, and it was the regular responsibility of some of the monks to tend to the guests' needs. The Le Grand Saint Bernard Hospice in the Swiss Alps—featuring the famous St. Bernard dogs—was founded specifically as an inn for travelers in A.D. 961 by Augustinian monks. This huge stone structure was one of the earliest identified inns and had up to 80 beds with the capacity of sheltering a total of 300 persons.

and motels. Bed and breakfasts offer overnight lodging to guests in a distinctive way. B & Bs are usually private homes, farmhouses, or other historic or period buildings with a small number of rooms. The most successful B & Bs are pleasant and homey creations, where the friendly touch of the owners makes each visit a personal and memorable experience for travelers. The Professional Association of Innkeepers International estimates that there are approximately 20,000 B & Bs in the United States—up from only 5,000 in 1980.

A third trend in the hospitality industry has been the construction of increasing numbers of retirement communities; as people live longer, and retire with more wealth, these self-contained communities offer private homes clustered around meeting, recreation, and dining facilities.

A fourth trend is the increase in the number of foreign tourists in the United States. According to the American Hotel & Lodging Association, the United States "receives a larger share of world international tourism receipts than any other country in the world." The number of international visitors to the United States increased by 10 percent from 2006 to 2007. The recent economic downturn has dampened this growth somewhat, but visitors from countries that have a good exchange rate with the American dollar continue to visit the United States in large numbers.

In the past, hotels and motels were set up primarily to serve the wealthy traveler or the person going from place to place. Today, with places such as Disney World, Universal Studios, Club Med, and others, many hotels are no longer simply places to stay while visiting a certain location. They have become travel destinations in themselves.

STRUCTURE OF THE INDUSTRY

There are four basic necessities of travel: transportation, lodging, dining, and entertainment. When planning a trip, most travelers first gather information on these four elements by checking the current availability and cost of transportation, lodging, and food at their planned destination. The entertainment aspects of the destination are considered as well. Some locations are interesting for their historical or cultural relevance, others provide pleasant natural surroundings, while yet others provide physical adventure. The tourist may choose to combine one or more elements in a package plan that someone else arranges, or the tourist may choose to create a vacation and handle all the planning personally. In both instances, travel specialists may be involved in providing guidance, information, and services to the traveler.

If the traveler is planning a trip without the aid of a travel specialist, or travel agent, the resources available are varied, depending on the destination. A number of written travel guides are available for almost every major destination, whether it's a city or a natural area, and travel guides for entire regions and countries are also abundant. There are hundreds of thousands on travel-related topics. Language books for travel abroad give translations and pronunciation guides for commonly used phrases. Audio files are also available for travelers who wish to learn a little of the local language before heading abroad. Classes, television shows, and an endless variety of DVDs, streaming videos, and books are available for foreign-language instruction specifically geared to the traveler. All the commonly studied languages are widely available and many of the more unfamiliar languages, such as African tribal languages, are also available.

Many travelers find it easier to let a professional help them with the information gathering and planning. In this case, they consult a travel agency. The goal of travel agents working for a travel agency is to help their clients plan a trip that meets their desires and fits within their travel budget. Specifically, they check rates on transportation and accommodations and make reservations. Agents also provide information such as visa and medical requirements for travel abroad, and they supply additional directions specific to the traveler's needs.

There is another option for travelers that simplifies the planning process even more: the packaged tour. Packaged tours—which can range from several days to several weeks—are available for those who wish to have many aspects of a trip planned in advance. They may cover a number of countries or they may stay in one city the entire time. Tourists have a wide variety of tours to choose from to meet their specific needs and interests. Travel agencies, private groups, museums, universities, and other institutions are just some of the organizations that provide package tours.

One specific type of packaged tour is adventure travel, which has become one of the fastest growing segments of the travel industry. This type of travel is geared toward the more physically active traveler who enjoys both seeing great wonders *and* exploring them. Adrenaline-pumping activities such as kayaking, whitewater rafting, and hiking are just some of the pursuits available to the adventure traveler. Hundreds of outfitters make a living organizing and guiding such trips, which may include a week-long trip sea-kayaking in Baja or whitewater rafting in the Grand Canyon. Outfitters usually specialize in one sport, but some of the bigger companies take on more than one. They usually take groups of eight or more people on their trips.

For tourists interested in spending much of the time actually moving from one point to another, cruise ships provide a slower, more leisurely type of travel. Cruise ships were the only form of travel across the oceans for many years. With the onset of air travel, cruise ships fell out of favor. Shorter cruises, in the Caribbean, for example, have once again gained in popularity. Cruise ships provide enough entertainment so that some passengers regard them as floating vacation spas. The locations visited by the ship may not be that important to the cruise passengers who choose whether or not to disembark at ports. Cruises usually run from three days to a few weeks and may dock in two or three cities. Some of the most popular cruises are to Alaska, Hawaii, and the Mediterranean/Greek Islands/Turkey.

Establishments in the hotel and motel industry fall into one of five categories: transient, motel or motor inns, residential, resort, and convention hotels.

Transient, also known as commercial, hotels make up three-fourths of the hotel business in the United States. These hotels cater to commercial travelers, businesspeople, salespeople, and tourists who wish to spend one or more nights at the hotel. Many commercial hotels have swimming pools, saunas, exercise rooms, ballrooms, conference rooms, and some house restaurants and drinking establishments open to the public as well as guests.

Motels are generally located near highways and airports and in small cities. Many motels offer parking beside or near the guest's room, eliminating any garage fees and unnecessary nightly loading and unloading of baggage. The facilities range from simply a room with a bathroom to motels with swimming pools and restaurants. Others have rooms that are designed as separate structures, each with a bed, bathroom, and kitchenette. These accommodations are referred to as suites. Some hotels offer suites only, mainly housing businesspeople on extended stays.

An inn is usually a small building, perhaps an extension of somebody's home, that provides simple services such as a clean bed and bathroom. The number of rooms range from five to 20. Bed and breakfasts are a growing branch of inns.

Residential hotels provide permanent or semipermanent housing, on a weekly, monthly, or sometimes yearly, payment basis. These facilities may offer amenities such as maid and food service. Some residential hotels may provide small kitchenettes in each room. This type of hotel has become increasingly popular in recent years—especially for business travelers who need to stay in an area for a long period of time, but not long enough to rent an apartment.

Hotels that offer recreational or social activities besides lodgings are considered resorts. Usually located near popular tourist attractions, resorts may have skiing, or during warmer weather, water activities, tennis, golf, or horseback riding, among others, to keep guests busy. Many of these hotels offer fine shopping and dining, themed décor, and entertainment. The resort as a complete family vacation destination is an apparent trend. Hawaii, Florida, Mexico, and the Caribbean are meccas of luxury resorts. Some resorts, especially those in Las Vegas and Atlantic City, are built around gambling activities.

Spas are similar to resorts, but differ in that usually everything required for the vacation is present in one location. If the guest decides he or she does not want to leave the grounds until the end of the vacation, the facilities provide everything from food to entertainment to keep the guests happy. The spas may choose to have an operating theme, such as physical health care, and many of the day's activities will focus on that. A health spa will include health-conscious menus, exercise classes, massages, relaxation therapy, and other aspects of health training and support.

Convention centers are usually a complex of buildings, one of which is a hotel. These centers are used as meeting places for large groups or businesses, or for major exhibitions; lodging for the conventioneers is provided by the adjoining hotel. Convention hotels and centers must have state-of-the-art audiovisual and technical equipment to stay competitive. Many of these convention centers are found in desirable, tourist-friendly locations, such as Las Vegas.

There are seven main branches of the hotel and motel industry. Front office, service, marketing and sales, and accounting comprise the "front of the house" positions, or those most visible to the public. Less visible "back of the house" jobs include food and beverage, housekeeping, and engineering and maintenance. Most branches of this industry operate on a three-shift system, allowing for 24-hour service for hotel guests.

The front office deals with all the paper and computer work involved with room and reservation assignments. The people working in this department also run the reservation desk, switchboard, and mail room. A general manager heads this department as well as the entire hotel operation. Department supervisors report to the general manager.

The main purpose of the service branch is to make the guests feel welcome. This includes greeting guests, parking cars, running the elevators, opening doors, carrying baggage, preparing a guest's room, and assisting with travel plans and entertainment. Most jobs in this department need little training or further education, thus creating a great starting place for employees eager to break into the hotel trade.

The accounting and financial management branch controls the fiscal affairs of the hotel. Projects such as financial policy and planning, maintenance of records and statements, overseeing expenditures, bank accounts, and payroll handling are some of the many responsibilities of this department. Though the accounting staff may not have one-on-one contact with hotel guests, this department is considered "front of the house" because of the

managerial nature of the work. Many accounting executives rise to leading hotel positions.

The marketing and sales sector strives to attract potential customers. Employees in this department try to find out what guests need and desire. Marketing and sales workers often use surveys, focus groups, or other research methods to gauge the feelings and opinions of guests and potential guests. The creative efforts of those in marketing and sales are expressed in new programs to attract guests or promotional campaigns designed to inform potential guests of current services.

The food and beverage department is among the largest and most lucrative sectors in the hospitality industry. It includes all the services involved with the bars and restaurants of a hotel, as well as room service, from purchasing and food preparation to presentation.

Depending on the size of the hotel, the housekeeping department can easily number in the hundreds. The room and floor attendants are responsible for keeping the rooms clean and supplied with fresh linens and towels. They also suggest repairs and improvements for guest rooms.

The engineering and maintenance departments keep the facilities of a hotel, motel, or similar establishment in working order. The chores include plumbing, painting, electrical wiring, and general repairing. They also help the housekeeping staff with the more physically demanding tasks of keeping a hotel clean.

CAREERS

We can break down the wide variety of careers available in the travel and hospitality industry by again keeping in mind the four basic necessities of travel—transportation, lodging, dining, and entertainment. A sampling of career opportunities in each area follows. The ultimate goal is the same for everyone in the industry: customer service and satisfaction.

Transportation (Trip Planning)

Before the fun or business of a trip can begin, it must be planned; travel and lodging reservations must be made, rental cars reserved, entertainment possibilities

explored. A variety of workers help people plan and organize their travel itineraries.

Reservation agents, for example, may work for hotels, airlines, health spas, train lines, cruise ships, or a number of other facilities directly related to one form of trip or another. They are responsible for checking the availability of accommodations and reserving a place for the customer. Agents confirm schedules, arrival times, departure times, and all the necessary information to ensure a smooth trip. Some agents can issue tickets, depending upon the tour companies they work for.

Like a reservation agent, the *ticket agent* is responsible for finding an open place for a customer, usually on a plane, and verifying the schedule of arrival, departure, and other necessary information. *Car rental agents* fill requests from travelers who require a vehicle while in the area.

For clients seeking specific assistance and information on certain types of travel, a *travel consultant* is available to help customize a vacation or travel plan. This assistance may include obtaining a visa for foreign travel, arranging special accommodations for the client, or adapting a schedule to fit the client's needs.

Serving some of the same functions as the travel consultant, the *travel agent* also issues tickets and reservations and may be able to design group packages and tours for a number of clients traveling together. Travel agents may specialize in one region or one form of transportation.

Travelers traveling by plane are taken care of during the flight by *flight attendants*. These workers, who spend most of their working hours in the air, help passengers board the plane, stow their luggage, and find seats. They also serve drinks and food during the trip, try to meet any specific needs a traveler may have, and ensure that passengers follow airline safely regulations.

Commercial airline pilots are those who actually fly the planes that carry passengers. They are responsible for rigidly following flight plans that dictate route, speed, and departure and arrival times. While en route, they monitor aircraft systems, watch weather conditions, and maintain constant communication with air traffic controllers. There are two main designations of commercial airline pilots: *captain* and *copilot*.

Tour guides sometimes accompany travelers on their trips. They provide assistance with accommodations, foreign languages, logistics of travel, and information about the places that will be visited. The tour guide is often responsible for arranging everything from transportation to and from airports to the menu to be served in the restaurants where the tour members will eat. In many locations, such as castles, museums, and historical sites, *group tour guides* present the place in an educational manner, conveying information that the tourist may not know.

Adventure travel outfitters plan and lead trips for physically active or sports-, culture-, or environment-minded groups of people to exotic places both in the United States and abroad. The successful outfitter will have a great deal of travel experience in the area in which he or she specializes. He or she usually has an office staff and a field staff of guides who actually lead the trips.

More Lingo to Learn

hospitality industry The industry that serves as an umbrella to the hotel and motel, or lodging, trade. As defined by the Council on Hotel, Restaurant, and Institutional Education, hospitality also includes food services, recreation services, and tourism.

niche marketing Hotels can attract a specific group by using special amenities and services to make foreign or special-needs guests feel at home. Some examples are internationalized service, multilingual staff, and audio or Braille hotel information and menus.

smart cards Also known as key cards, these are electronically coded plastic cards about the same size and width as a credit card. These key cards are inserted into slots located near the guest room; if the information matches that stored at the front desk the room door will open. Cards are recoded with every new guest.

Travel writers help people plan for trips by writing about travel and the hospitality and tourism industries for newspapers, magazines, books, and Web sites (including blogs). They also appear on radio and television broadcasts to talk about travel-related topics.

Lodging

Chances are you have stayed at one of the more than 48,000 hotels and motels found throughout the United States. These places, from the grand luxury of the Waldorf Astoria, to the Bates Motel, to the more ubiquitous HoJo, may differ in levels of accommodations (not to mention room service, or lack thereof), but all do share the same basic purpose of providing safe and comfortable lodging for travelers.

When people travel away from home they obviously need a comfortable place to stay and relax. All hotel and motel centers need conscientious, well-trained employees. Since customer contact is a rudimentary and essential element of any hotel industry position, it is helpful for employees to possess good communication skills, patience, a diplomatic demeanor, and a readiness to help.

By dividing the work of a hotel into front-of-the-house, where employees are highly visible to the guests, and back-of-the-house, where employees generally work behind the scenes, we can get a better understanding of the various opportunities available in the hotel and motel industry.

Front-of-the-House

Doorkeepers and *baggage porters*, also known as *bellhops*, help guests as they arrive at or depart from the hotel. They often direct customers to the check-in counter and then usher them to their rooms. Bellhops sometimes assist guests in hailing taxis. *Bell captains* supervise bellhops and doorkeepers.

Administrative positions, such as *desk* and *reservation clerks*, account for 15 percent of jobs in the lodging industry. Their many duties include coordinating reservations and room assignments, greeting guests, furnishing room keys or key cards, and forwarding mail and any messages. Front office workers also handle complaints and are often the employees who guests turn to most frequently for assistance.

The position of the *concierge* started as a European custom, but now is a staple in many larger urban hotels. Among their many duties, the concierge can handle requests for special services such as reserving theater or sports tickets. Since a concierge may be asked for recommendations to restaurants, museums, or other entertainment options, he or she must be well versed in the city's cultural and tourist attractions. Most concierges are multilingual to better serve international guests.

Each area of the hotel—from the restaurant to security to the front desk—requires *managers*. *Restaurant managers* oversee the daily operations of the hotel's restaurant or restaurants. They report to the *food and beverage* manager, who is responsible for all food and beverage services in the hotel. (See the Dining section below for more information on these types of managers.) A *security manager*, sometimes known as a *director of hotel security*, is entrusted with the protection of the guests, workers, and grounds and property of the hotel. *Resident managers* live in the hotel and are on-call twenty-four hours a day in case an emergency or other situation requires immediate action. They supervise all the hotel departments, as well as handle guest complaints. *Front office managers* supervise all front office personnel and activities, including scheduling work assignments, creating and managing reservations systems, and overseeing guest relations. *General managers* supervise the overall operations of a hotel establishment. They coordinate front desk service, bell service, housekeeping, and other aspects of day-to-day operations.

The financial department oversees all money that comes into a hotel. It is responsible for recording sales, controlling expenditures, and keeping track of overall profits or losses. Financial employees include *controllers, head accountants, cost accountants, credit managers, accounts receivable supervisors, accounts payable supervisors,* and *bookkeepers.*

Sales and marketing staffs are part of the larger hotels and franchise operations. The *director of marketing and sales* oversees a staff of sales managers and sales and marketing workers. As *public relations specialists*, sales and marketing staff constantly strive to establish and preserve the hotel's reputation before the public by arranging for favorable radio, television, newspaper, and magazine publicity and by seeking additional business clientele. Sales and marketing workers also often conduct surveys designed to monitor the

> ### Quote
>
> "If you reject the food, ignore the customs, fear the religion, and avoid the people, you might better stay home."
>
> — James Michener, American author

feelings and opinions of guests or potential guests.

The human resources or personnel department is responsible for hiring and firing the employees of a hotel. They also make sure that employees are productive and happy in their duties. The *personnel manager* heads the department. Larger hotels employ *training managers*, who oversee the hotel's management training program. Other employees in this department include *benefits coordinators*, who handle employee benefits such as health insurance and pension plans, and *employee relations managers*, who deal with employee rights and grievances with an overall goal of creating a positive and productive work atmosphere.

Back-of-the-House

The housekeeping staff is responsible for keeping guestrooms and the rest of the hotel clean and orderly. *Housekeepers*, sometimes known as *maids*, keep hotel rooms clean. They also inspect rooms, stock linens and toiletries, and provide additional sundries or other services, such as picking up and delivering dry cleaning for guests. In large hotels, *executive housekeepers* may supervise housekeepers and other personnel.

Maintenance and engineering workers maintain and repair the facility's equipment. They keep the electrical wiring and appliances, the plumbing, and the numerous machines in good condition and make sure that they are working smoothly. *Janitors* also clean the premises to attract customers, to improve safety, and to reduce wear and tear on the establishment.

Dining

In vacation settings such as spas, cruise ships, cross-country trains, and hotels and motels, food is provided to the tourists while they are in residence. The food service staff may have different titles in different locations, but essentially function as *cooks, bakers, waiters* and *waitresses*, and *hosts* and *hostesses*.

Restaurant managers and *food and beverage managers* direct the activities in an establishment's cocktail lounge, restaurant, and banquet facilities. Restaurant managers hire, train, and supervise chefs, food preparers, waiters and waitresses, dishwashers, wine stewards, buspeople, and bartenders. They deal with suppliers, make sure inventories are at their proper level, and submit daily statements on restaurant sales to the food and beverage manager. The food and beverage manager is responsible for all food and beverage services in the hotel. He or she makes decisions regarding food purchases, kitchen equipment, facility décor, and employee uniforms. They also work with restaurant and other facility managers to determine menu selections and prices.

Hotel convention service coordinators, sometimes known as *event planners,* plan and organize hotel events such as meetings, trade shows, musical performances, and wedding receptions. They work with customers to make an event successful. This may include responsibilities like ordering flowers, arranging a menu, hiring a musical group, and making sure the convention area is clean and presentable for the occasion.

Entertainment

Spas and cruise ships also often have *entertainment directors,* who plan parties and other events, and often act as hosts or hostesses. On a cruise ship, the *cruise director* oversees the staff that provides all entertainment on the ship. The entertainment staff includes *performers, movie theater workers, fitness instructors,* and any other employee who provides service to the passengers.

On large cruise ships, *cruise hosts* and *hostesses* handle many of the day-to-day aspects of directing the passengers to and from their entertainment activities. Some cruise lines employ hosts and hostesses to serve as dance partners or escorts to their passengers traveling alone.

Spa directors design the theme and activities of spas and hire staff. They may be the owners of the facilities as well.

EMPLOYMENT OPPORTUNITIES

The travel and hospitality industry is an important segment of society, employing millions of people and providing billions

Hotel Workers by Ethnicity, 2004

Hourly Employees

White:	39 percent
Hispanic American:	28.5 percent
African American:	23.8 percent
Asian American:	6.6 percent
Native American:	1.0 percent
Other:	0.7 percent
Pacific Islander:	0.5 percent

Line-Level Supervisors

White:	41 percent
Hispanic American:	28.2 percent
African American:	20.3 percent
Asian American:	6.7 percent
Native American:	2.4 percent
Other:	0.8 percent
Pacific Islander:	0.6 percent

Salaried Managers

White:	71.7 percent
African American:	10.2 percent
Hispanic American:	9.9 percent
Asian American:	6.5 percent
Other:	0.7 percent
Native American:	0.5 percent
Pacific Islander:	0.5 percent

Source: "Turnover and Diversity in the Lodging Industry," American Hotel & Lodging Association and the Hospitality Industry Diversity Institute

There are opportunities in every region of the country for those who are interested in the travel branch of this industry. Tour guides and tour managers work for tour operators. Employment with tour operators may begin with a listing of major travel companies advertising for tour guides and/or managers. This list can be obtained at the office of a local travel agent. As with other travel careers, guides and managers are advised to undertake training to hone their skills and increase productivity. Individuals who prefer to work solo within the industry may choose to go into a freelance career as either a travel specialist who advises travelers on specifics of a tour or an adventure travel outfitter.

Major airlines, such as American, United, Delta, Southwest, and U.S. Airways, employ thousands of workers in various capacities. In addition, there are a number of smaller, more regional airlines that likewise offer a wide variety of employment possibilities. Rental car companies are also big employers in the travel and hospitality industry, with major companies such as Avis, Hertz, Budget, Enterprise, and Alamo operating all over the country. There are several cruise lines, all of which hire employees to work in various capacities; Cunard, Princess, and Carnival are examples of major cruise lines. Additionally, The National Railroad Passenger Corporation (Amtrak), which serves more than 500 destinations in 46 states and the District of Columbia, offers many employment opportunities for food and cleaning service workers, baggage porters, engineers, conductors, and others.

of dollars each year to the national economy. Acquiring a job within the industry depends on the job seeker's expectations, experience, qualifications, goals, and ambition.

Most job opportunities in the hospitality branch of this industry are in highly populated areas or traditional tourist resorts, where hotel rooms are in higher demand. Lodgings in smaller towns or out-of-the-way places tend to be family-run establishments.

People interested in hotel careers may find employment with such large hotel and motel chains as Best Western International, Hilton Hotels Corporation, Global Hyatt, Wyndham Worldwide, and many others. These chains may offer the best opportunities for positions in management, sales, marketing, and public relations. Some chains, such as Marriott International Inc., offer "fast track" management programs designed to encourage career advancement for women and minorities, along with day care and other family-friendly benefits.

INDUSTRY OUTLOOK

As one of the largest retail service industries in the United States, the travel and tourism industry offers tens of millions of jobs. Overall employment prospects in the industry should be good over the next decade, but this industry is strongly affected by economic conditions. When economic conditions are poor, fewer people travel (and stay at hotels) for business or pleasure. When economic conditions are strong, hotels are booked; planes, trains, and cruise ships are filled with passengers; and cities bustle with tourists.

Tight competition among large and small travel companies will result in industry consolidation as the larger companies dominate the smaller. Other changes are also affecting the way the industry is structured. People are taking shorter but more frequent vacations. Three- and four-day weekends are replacing the two-week vacations that were common in the 1960s and 1970s. As the working population in the United States finds it more difficult to take extended vacations, many travel organizations have had to structure their programs around shorter trips. In addition to the growth in adventure travel, another trend is the growing popularity of ecotourism, which involves visiting a pristine natural area, learning about its ecosystem, perhaps even performing some work while there, and making every effort to preserve and protect that ecosystem without altering it by the act of traveling there. Ecotours to such places as the Galapagos Islands and Costa Rica have become very popular.

Yet another very strong and growing segment of the field is the cruise industry. People are opting more and more often for cruises, which offer a total vacation for a set price, and typically include several land tours in various ports of call. About 12.6 million people took a cruise in 2007, according to Cruise Lines International Association, and the U.S. cruise industry generated nearly 348,000 jobs.

While adventure travel, ecotours, and cruises are gaining in popularity, the traditional package tour is losing ground. This may be due in part to how much easier travel planning has become as a result of the Internet. Since a computer user can now sit at home, access almost all the necessary information about any given location, and basically plan his or her own package tour, the demand for others to do this is lessening. This same easy access, in addition to the increasing popularity of making actual reservations online, may ultimately affect travel agencies as well. Attempting to target the specific needs of a specific clientele has proven to be effective for many travel companies (with the business traveler who works for a small firm as the main target). To compete with the larger travel agencies, local and regional agencies use the focus approach to attract these small businesses. This approach is expected to result in additional careers in the industry, namely in marketing and sales.

International tourism has surged with the successful promotion of American destinations abroad and the relatively weak dollar. With large numbers of international tourists coming to the United States, hotels must learn to "internationalize" service, from international cuisine to multilingual staff, room directories, and information. Employees familiar with Asian languages and cultures will especially be in demand.

More establishments will target the fast-growing elderly population, many of whom have substantial retirement funds. Called assisted living communities, these complexes will offer housing, food, and medical services.

Consolidation will be key to growth in the future. Consortiums composed of smaller hotels and motels will pool resources and share advertising costs to edge off competition from bigger and better-known names such as Marriott and Hilton. Many larger hotels have joined with airlines, car rental agencies, and travel agents to offer complete travel packages. These businesses can provide savings and convenience for travelers while increasing their name recognition and improving their business.

It has been predicted that this field will employ more people than any other industry by 2020. This industry also employs a high percentage of women and minorities compared to other industries. New facilities are being built, many in popular travel destinations like Las Vegas and Orlando. Though many employees in the hotel industry can advance from unskilled, entry-level positions, promotion opportunities will be best for people with college degrees in hotel or hospitality management. It will become increasingly important to recruit skilled workers to fill new openings. The problems of finding and keeping staff are expected to become more difficult as the pool of younger workers shrinks.

The entire travel and hospitality industry is sensitive to political crises, such as terrorist acts or civil wars, and shifts in economic conditions both here and abroad. Consequently, the number of overall job opportunities fluctuates and is hard to predict. When the U.S. dollar rises, more Americans travel abroad since they are able to buy more with their dollar. Foreigners are less likely to travel to the United States when the dollar is strong because they are forced to spend more of their currency. Despite fluctuations in the market, it is likely that the travel and hospitality industry will remain strong for many years to come. It remains the single most important industry in many cities and regions.

SECTION 2

Careers

Adventure-Travel Specialists

SUMMARY

Definition
Adventure-travel specialists develop, plan, and lead people on tours of places and activities that are unfamiliar to them. Most adventure travel trips involve physical participation and/or a form of environmental or cultural education.

Alternative Job Titles
Adventure outfitters
Adventure travel guide
Adventure travel planners
Ecotourism specialists

Operations managers
Outfitters
Trip planners

Salary Range
$17,2200 to $30,360 to $75,000+

Educational Requirements
High school diploma; bachelor's degree required for some jobs

Certification or Licensing
None available

Employment Outlook
About as fast as the average

High School Subjects
Anthropology and archaeology
Earth science
Geology
Geography/social studies
Sociology

Personal Interests
Camping/hiking
The environment
Exercise/personal fitness
Helping people: personal service
Travel

"Although I had traveled extensively, when I went to Nepal, it was the most unusual culture I had seen," says Dr. Antonia Neubauer, when asked what prompted her to start her adventure travel company, Myths and Mountains Inc. "I couldn't speak the language, and sat on gods, thinking they were rocks! I came back, found a Nepali student at a local college who had just shrunk all her clothes in the washer, and told her I would teach her to use the washing machine if she would teach me Nepali. She did and I did.

"Subsequently," continues Dr. Neubauer, "I went back to Nepal, then people asked to go with me, and then they asked, 'Where are we going next?' Out of this came Myths and Mountains. Basically, it was a question of the 'right thing at the right time.' In a sense, if you take people on a trip, you teach a class. If you design a program, you design curriculum. If you run a school, you run a company, and if you do nonprofit work, you can start a nonprofit. Thus, Myths and Mountains was simply taking all I had done earlier in my life and transferring it to the field of travel."

WHAT DOES AN ADVENTURE-TRAVEL SPECIALIST DO?

Depending upon where he or she works, an average day for an adventure-travel specialist might be anything from planning tours in the comfort of an air-conditioned office to leading a safari through southern Africa or a cultural expedition through Mongolia.

Adventure travel is one of the fastest-growing areas of specialization within the travel industry. *Adventure-travel specialists* plan—and may lead—tours of unusual, exotic, remote, or wilderness locations. Almost all adventure travel involves some physical activity that takes place outdoors. Sometimes adventure travel is split into two different categories: soft adventure and hard adventure. Hard adventure requires a fairly high degree of commitment from participants, as well as advanced skills. A hard-adventure traveler might choose to climb Yosemite's El Capitan, raft the Talkeetna River in Alaska, or mountain bike through the logging trails in the Columbia River Gorge. Soft-adventure travel, on the other hand, requires much less physical ability and is usually suitable for families. Examples of this kind of travel might be a guided horseback

---

> ## Lingo to Learn
>
> **adventure travel** Travel away from one's local environment that includes activities adventurous to the participant.
>
> **ecotourism** Travel to natural areas that conserves the environment and improves the welfare of local people.
>
> **hard adventure** Activities with high levels of risk that require intense commitment and advanced skills.
>
> **nature-based tourism** Travel away from one's local environment that includes interaction with a natural environment for education, observation, or recreation.
>
> **soft adventure** Activities with a perceived risk, but low levels of actual risk, requiring minimal commitment and beginning skills; most of these activities are led by experienced guides.

---

ride through the Rocky Mountains, a Costa Rican wildlife-viewing tour, or a hot-air-balloon ride over Napa Valley, California. Other types of soft adventure might include trips to explore the culture and natural environments of a country or region.

Adventure-travel specialists are the travel professionals who plan, develop, and lead these types of tours. Some work strictly in an office environment, planning trip itineraries; making reservations for transportation, activities, and lodging; and selling tours to travelers. Others, typically called *outfitters*, work in the field, overseeing the travelers and guiding the tour activities. In some cases, the adventure specialist both plans the logistics of the trip and guides it.

For every adventure tour that takes place, numerous plans must be made. Travelers who purchase a tour package expect to have every arrangement handled for them, from the time they arrive at the starting point of the trip. That means that ground transportation, accommodations, and dining must all be planned and reserved. Each day's activities must also be planned in advance, and arrangements need to be made with adventure outfitters to supply equipment and guides.

The *trip planner* calls lodges, hotels, or campgrounds to make reservations for the tour group and arranges ground transportation, which may be vans, buses, or jeeps, depending upon the particular trip. He or she also works with the adventure outfitters who will actually lead the tour group through the planned activities and supply the necessary equipment.

Some companies serve as adventure-travel brokers, selling both tours that they have developed and tours that have already been packaged by another company. Travel specialists are responsible for marketing and selling these tours. They give potential customers information about the trips offered, usually over the phone or via email. When a customer decides to purchase a tour package, the travel specialist takes the reservation and completes any necessary paperwork. Depending upon their position in the company and their level of responsibility, adventure travel planners

may decide where and how to advertise their tours.

Working as an adventure travel out-fitter or *guide* is very different from working as an adventure travel planner or broker. The duties for these individuals vary enormously, depending upon the type of tours they lead. Adventure tours can take place on land, on water, or in the air. On a land adventure trip, guides may take their tour groups rock climbing, caving, mountain biking, wilderness hiking, horseback riding, or wildlife viewing. They may take them to a variety of cultural sites such as the World Heritage site of Luang Prabang in Laos, Etosha National Park in Namibia, or Machu Picchu or the sandy ruins of Chan Chan in Peru. On a water trip, they may go snorkeling, scuba diving, surfing, kayaking, whitewater rafting, or canoeing. Air adventures include sky-diving, parasailing, hang gliding, bungee jumping, and hot air ballooning.

Whatever the nature of the trip, guides are responsible for overseeing the group members' activities and ensuring their safety. They may demonstrate activities, help with equipment, or assist a group member who is having difficulty. In many cases, where travelers are interested in the scenery, geography, local peoples, wildlife, or history of a location, guides serve as commentators, explaining the unique aspects of the region as the group travels.

Guides are also responsible for helping tour group members in the case of an emergency or unplanned event. Depending upon the nature of their tour, they

> ## To Be a Successful Adventure-Travel Specialist, You Should...
>
> - enjoy spending time outdoors (guides)
> - have good business and math skills (office workers)
> - be responsible and able to think on your feet (guides)
> - be a good writer (office workers)
> - be detail-oriented
> - communicate well with others
> - be in good physical condition (guides)
> - know emergency first aid and CPR (guides)

must be prepared to deal with injuries, dangerous situations, and unusual and unplanned happenings. Essentially, it is the guide's responsibility to ensure that tour group members have a safe, memorable, and enjoyable trip.

WHAT IS IT LIKE TO BE AN ADVENTURE-TRAVEL SPECIALIST?

Karen Cleary is the operations manager at Boundless Journeys, a small group adventure tour operator in Stowe, Vermont, that was named one of the "Best Adventure Travel Companies on Earth" in 2009 by *National Geographic Adven-*

ture magazine. "We have a small office staff and work with a network of local guides and partners all over the world to organize active travel experiences for mostly North American clients," she says. "Having an affinity for the outdoors and a love of travel, as well as being one of those ultra-organized personalities, planning high-quality travel experiences seemed like a perfect fit for me.

"Undeniably," she continues, "the biggest pro is traveling for work—and the type of travel is amazing. I have researched trekking trips in Peru; safaris in Botswana, Zambia, and Tanzania; and walking tours in Ireland, Iceland, and Costa Rica. But there are other pros: the work is interesting and varies day by day. Some days I am editing copy for a new catalog, or negotiating contract terms with a local transportation provider, or coaching a local guide on how to handle a challenging guest, or simply dreaming up a great new trip. The con, if this is one, is that this is not a profession to get into for the huge monetary rewards. The salaries for travel professionals really vary, and increases do come with experience, but this job is much more about the rewards I mentioned earlier than the paycheck. Also, with worldwide travel, a crisis can come up at any time, and now and then the phone at home rings at 6 A.M. or 10 P.M. with a problem that needs immediate attention."

One of Karen's most memorable experiences working in the field was the opportunity to bring her mother to Botswana and Zambia last year on a research trip through the Okavango Delta. "It was a once-in-a-lifetime experience," she says, "and being able to share it with her was wonderful."

Dr. Antonia Neubauer is the president and founder of Myths and Mountains Inc., an adventure travel company in Incline Village, Nevada. In 2009 *National Geographic Adventure* magazine voted it one of the "Top 10 Best Adventure Travel Companies on Earth." She also is the founder of READ Global, a nonprofit global organization dedicated to making rural villages viable places to live. READ builds library community centers, seeds businesses to fully sustain and support the libraries, and then links the library community centers with organizations providing needed village services— health, literacy, microcredit, etc. (See the sidebar "Inside READ Global" for more on this organization.)

"As head of a small company," Dr. Neubauer explains, "my major duties are providing leadership and direction for the organization; general administration; financial management and oversight; staff hiring, management, and oversight; being a general spokesperson for the company; corporate oversight; looking at new sites or new directions in which to move; sales, particularly Asia and Southeast Asia; and serving as chief cook and bottle washer. We are small, so we all have to collaborate, and my job is making sure that we have a collaborative environment in which to work."

Dr. Neubauer says, "There is a huge difference between a typical day in the office and an excursion. People like to think that travel is romantic, but most is very routine. In the office, the bulk of a typical day is spent answering email inquiries, talking with prospective and booked clients, writing and pricing itineraries, and handling email correspondence with overseas operators.

"From a managerial perspective," she continues, "there are meetings and meetings and meetings—regular meetings with staff and any clients who may come to visit, outside sales calls or marketing sessions, or discussions with business advisers. Often we give presentations at public meetings, conventions, or conferences about travel, a particular country, or another subject for different organizations.

"On an excursion, if you are leading a trip, you are there for everyone else

Ecotourism: Traveling Responsibly

Ecotourism is a fast-growing segment of the travel and tourism market that is often associated with adventure travel. This type of travel is defined by the International Ecotourism Society as "responsible travel to natural areas which conserves the environment and improves the welfare of local people."

Ecotourism activities emphasize the goal of preserving the natural areas that tourists are visiting. They often combine outdoor recreation with learning about a region's natural history and ecology. Some ecotours even include seminars on historic or wildlife preservation or community-service projects.

in the trip. You get up early and make sure everyone is feeling well, had a good night's sleep, have all their things ready to go, and answer any questions that come up. Most importantly, you make sure everyone knows the schedule for the day and has what is needed for touring. As you travel, you create the very special situations in which your travelers learn about a country, provide them with appropriate educational background information, interface with all of the operating staff and make sure they are happy, watch to see that no one gets lost and everyone is both going at their own pace and yet functioning as part of a group. If you are lucky, you have time to eat a meal while helping both your travelers and staff. Hopefully you do not get sick, because if you do, there is no time for that and you need to be there for your group. If someone is hurt, it is your job to make sure what needs to happen medically does happen."

DO I HAVE WHAT IT TAKES TO BE AN ADVENTURE-TRAVEL SPECIALIST?

A love of the outdoors is perhaps the most important characteristic of travel specialists who work in the field. It's also important for adventure travel tour guides to have a passion for sharing their love of nature and their knowledge with others.

An educational background in the natural sciences or cultural anthropology is important for some adventure travel guides; for others, a high skill level

in certain activities, such as rock climbing or cross-country skiing, is necessary. Whatever type of tours you guide, however, being in reasonably good physical shape is a must.

Being mature and responsible is important in this sort of job, where you are leading groups of people through areas and activities unfamiliar to them. Guides should be trained and confident in performing emergency first aid and CPR.

Adventure travel professionals who work in an office, developing and selling tours, need some different personal qualities than those who work in the field. You need to have a good work ethic, be friendly and confident, and have good phone skills. While being an active, outdoorsy person may help you sell tours, it is not a requirement for working in this branch of adventure travel. Attention to detail and good organization skills are more important.

Karen Cleary says that the most important personal and professional skills for adventure-travel specialists are "creativity, enthusiasm for travel, obsessive organization, cultural sensitivity, typical computer skills (word processing, spreadsheet applications), math (for calculating profit margins, etc.), statistics/business analysis skills (for management-level jobs), and, lest we forget that these are for-profit businesses, an ability to sell to a discerning, well-educated clientele."

Dr. Neubauer says that key professional qualities for workers in the field are "the ability to write and speak well, good people skills and love of people, destination knowledge for sales or leading a trip, management and personnel skills, a vision and sense of values, creativity and passion for what you are doing, computer skills, good judgment, professionalism, attention to detail and desire for excellence, and foreign travel experience and language skills."

HOW DO I BECOME AN ADVENTURE-TRAVEL SPECIALIST?

Education

High School

If you are considering the business end of travel—working in a brokerage, planning tours, or eventually owning your own tour-packaging business—you should start taking business courses while still in high school. Accounting, computer science, mathematics, or any other business-related course will give you a good start. Classes in geography, geology, social studies, and history might also help you understand and discuss the locations with which you may be dealing. Finally, classes in English or speech are always good choices for helping you develop the ability and confidence to deal with people.

If you are more interested in the fieldwork aspect of adventure travel, you will need to take classes that help you understand how the earth's environment and ecosystem work. Because tour guides often explain the natural history of a location, or educate tour groups on local wildlife and plant life, classes in earth science, biology, and geology are excellent choices. Classes that teach you about the social history of various places—such as social studies or anthropology—might also be beneficial.

Since much adventure travel involves physical activity, which may range from low- to high-impact, taking courses or becoming involved in activities that promote physical fitness is a good idea. If you already have an interest in a particular area of adventure travel, you may be able to join clubs or take classes that

---

Good Advice
● ● ● ● ● ● ●

Dr. Antonia Neubauer offers the following advice to high school students who are interested in entering the field of adventure travel:

- Go places, see people, do things.

- Get a good, solid education (business/finance/accounting, leadership, administration, foreign languages, and marketing—both regular and e-marketing) in college and beyond.

- Know that there is a lot of diddly day-to-day work in every business, and don't just think of the "romantic fairy tale" world of travel.

- Make sure you have good writing and speaking skills—a rarity today.

- Make sure you are in good physical shape—travel is not easy in our business.

- Be creative and think out of the box.

- If you can't laugh, find another job.

Good Advice

Karen Cleary offers the following advice to high school students who are interested in entering the field of adventure travel:

- Learn geography and world history so that you are well versed in adventure/cultural travel destinations.

- Read about places that interest you. There are literally thousands of exciting books covering everything from Antarctic exploration to China's Cultural Revolution that will broaden your world perspective.

- Learn to write well. Much of what we sell is done through evocative writing; it has to entice people and be error free.

- Travel. Consider a semester abroad or summer travels while in college. Intern at a tour company if there is one in your area, or spend a summer guiding adventure outings if you live in an area known for its recreation (rafting, sea kayaking, and mountain biking/cycling shops often offer local tours that you may be able to help guide). Cycling-oriented adventure travel companies in particular seem to hire younger guides with bike maintenance skills rather than Ph.D. naturalists or anthropologists. This is a great way to gain experience and spend a summer exploring a new place. Note that you'll probably need first aid/CPR credentials to do this.

- Learn a second language, or bits of several. Even if you are not fluent, everyday pleasantries are valued by local people.

help you develop the right skills. For example, scuba diving, sailing, hiking, mountain biking, canoeing, and fishing are all adventure travel activities that you might be able to engage in while still in high school.

Postsecondary Training

There are several different approaches you can take to prepare for a career in adventure travel. While it may not be necessary for all jobs, a college degree will likely give you a competitive edge in most employment situations. If you choose to obtain a college degree, some options for majors might be earth science, biology, geology, world history, natural history, or environmental affairs. If you hope to become involved with an intensely physical form of adventure travel, a degree in health, physical education, or recreation may be a good choice.

If you are more interested in the planning and reservations end of adventure travel, a college degree in business or marketing is a good choice. Some adventure travel brokers suggest that attending one of the many travel agent schools also provides a good background for the administrative aspects of the business.

It may be possible to find a job in adventure travel without college training if you happen to be very experienced and skilled in some form of adventure activ-

Likes and Dislikes

● ● ● ● ● ● ● ● ● ●

Dr. Antonia Neubauer details what she likes most and least about work in adventure travel:

Pros

- Travel to wondrous places

- Friends all around the world

- In my case, the chance to be very creative—one never stops learning

- The opportunity to change the lives of both traveler and local in a positive way forever

- A sense of the world as a whole, not a narrow parochial view of life

- Incredible and unlikely teachers from all over and all walks of life

Cons

- Not a business to "get rich with"

- A lot of hard work

- The most difficult thing is always dealing with people issues

- You are subject to economic and political vagaries that can impact greatly on business

- Changing business with the Internet—this can be a con or a plus, but it is not easy to manage

ity. If you choose this path, you should spend as much time as possible developing whatever skill you are interested in. There are classes, clubs, and groups that can teach you anything from beginning diving to advanced rock climbing.

Certification or Licensing

No certification or licensing is currently available for this profession.

Internships and Volunteerships

If you attend college, you will most likely be required to participate in an internship that helps you explore your field of interest. Participating in internships is an excellent way to try out careers and meet others who have similar interests, as well as potential employers. You can also try volunteering with a local tour operator during your summer vacation or after school to get an introduction to the industry.

WHO WILL HIRE ME?

In the last decade, there has been an enormous increase in the number of adventure travel providers. In addition to this growth in commercial suppliers, a number of not-for-profit organizations—such as universities and environmental groups—are also offering nature and adventure programs.

Your first step in finding a job should be to develop a list of American and Canadian adventure travel wholesalers and outfitters; you might talk to a travel agent or check with the reference desk of your local library. Another option is to get on the Web and perform a keyword search on "adventure travel" or "outfitters." Many of these organizations have their own Web sites. Professional associations, such as the Adventure Travel Trade Association (http://www.adventuretravel.biz) and the U.S. Travel

Inside READ Global

● ● ● ● ● ● ● ● ● ●

Dr. Antonia Neubauer provides more information on READ Global, a nonprofit organization she founded in 1991:

When we created Myths and Mountains, we started a nonprofit to give something back to the people in the lands in which we had worked. The goal was to create something that was holistic, not a Band-Aid on a bleeding artery, and something that created independence from foreign aid or charity. Moreover, having worked in education and traveled extensively, I had seen so many of the failures of philanthropy, so many do-gooders who really caused more problems than they solved, and so much foreign aid that was wasted.

Thus, the goal of READ became to make villages viable places for people to live, learn, and prosper. READ inspires rural prosperity by building a rural library/community center and seeding a local business (ambulance service, furniture factory, mill, catering operation, storefront rental, etc.) whose profits fully sustain and support the library. Then READ links the library community centers with other organizations providing needed village services—health and HIV [screenings/treatments], microcredit, literacy, agriculture or livestock classes, etc.

Each of the library/community centers has 3,000-5,000 books in the local language, a women's section, an early childhood section, a regular reading room, an audio-visual section, a computer section, and a cultural section. READ trains local people to serve as librarians. The sustaining businesses are selected by the villagers and provide, on average, at least three to five jobs per community. The libraries belong to the village and are run totally by local committees—management committee, women's committee, education committee, finance committee, and student committee.

There are several other key parts to the READ program:

- Villages approach READ. READ does not go to them. Villagers have to write a proposal themselves. If they do not care to do this, they will not take care of a library.

- Villagers put in 15–20 percent of the costs of the library. In many cases, they even contribute more than READ.

- Villagers have to donate the land for the library.

- Villagers are in charge of the process, monitored by the READ country office.

- All READ staff in a country are native to that country.

- This is a true bottoms-up, asset-based approach to development.

- All libraries must be totally self-sustaining with money generated from the local businesses.

In 2006 READ Nepal won the $1 million Bill and Melinda Gates Foundation's Access to Learning Award. Then in 2007, READ Global received a $3 million replication grant from the Gates Foundation. To date, READ Global has expanded into India and Bhutan, as well as Nepal. The goal is to expand into four other countries in the next three years.

Association (http://www.tia.org), also offer job listings at their Web sites.

There are a number of magazines that may be helpful in compiling a list. Some good publications to look into are *Outside*, *Backpacker*, and *Bicycling*. *National Geographic Adventure* publishes an annual list of the best adventure travel companies. Visit its Web site (http://adventure.nationalgeographic.com) for more information. A final method of getting a list of travel wholesalers and outfitters is to contact one or all of the adventure travel organizations listed at the end of this book. These associations should be able to give you a list of their members.

To find not-for-profit organizations that hire adventure-travel specialists, consider the National Audubon Society (http://www.audubon.org), Earthwatch Institute (http://www.earthwatch.org), and the Sierra Club (http://www.sierraclub.org). Again, check with your local library for a more complete listing of environmental groups. You might also contact universities to see if they have a wilderness/adventure travel or outdoor recreation division in their schools of physical education or recreation.

Once you have a list of adventure travel companies, you may need to do some research to discover what sort of activities they provide. That way, you can narrow your search to companies that specialize in the activity or activities you are experienced in. Remember that for your best chance of finding a job in adventure travel, you may have to relocate, so your search should be geographically broad. After you've chosen the companies to which you will apply, you should either send a resume and cover letter directly to each company or make a preliminary phone call to inquire about possibilities.

You should also use any contacts you have—from clubs, organizations, previous travel experiences, or college classes—to find out about possible employment opportunities. If you belong to a diving or bicycling club, for example, be sure to ask other members or instructors if they are familiar with any outfitters you could contact. If you have dealt with outfitters in some of your own adventure trips, you might contact them for potential job leads.

WHERE CAN I GO FROM HERE?

There is no clearly defined career path for adventure-travel specialists. For those who work in an office environment, advancement will likely take the form of increased responsibility and higher pay. Assuming a managerial role or moving on to a larger company are other advancement possibilities.

For those who work in the field, advancement might mean taking more trips per year. Adventure travel in many locations is seasonal, so tour guides may not be able to do this sort of work year-round. It is not uncommon for an individual to guide tours only part time and have another job to fill in the slow times. If a tour guide were to become experienced in two or more particular areas of travel, however, he or she might be able to spend more, or even all, of the year doing adventure touring.

Another option for either the office worker or the guide would be to learn about the other side of the business.

Adventure Travel Company Operator Spotlight: Mary Dell Lucas

● ● ● ● ● ● ● ● ●

Mary Dell Lucas is the founding director of Far Horizons Archaeological and Cultural Trips Inc., which is headquartered in San Anselmo, California. (Visit http://www.far-horizons.com to learn more her company and the tours it offers.) She spoke with the editors of *What Can I Do Now? Travel and Tourism* about the field.

Q. Please tell us about Far Horizons and your background. Why did you decide to enter this career?

A. I was an archaeologist working in the field excavating and going to school to get a master's and Ph.D. I was a returning student; I had been a flight attendant for more than a decade before returning to school. Part of the way through my MA, it became clear to me that I really didn't want to end up with a Ph.D. and no job, but I still wanted to keep my connection with archaeology. I created Far Horizons Archaeological and Cultural Trips more than 25 years ago.

Q. What are your main and secondary job duties?

A. Each day is different. In the office, we design future tours, handle the logistics of upcoming trips, and promote the business through email, mailings, and advertising. When accompanying a tour group, the tour manager is responsible for the daily logistics, and

to ensure that the participants are healthy and happy.

Q. What are some of the pros and cons of work in this field?

A. You're asking this question at a bad time since the economic downturn is hurting the travel business very badly. The good side of the job is that there is never a dull moment and every day is different. And of course, the ability to travel to foreign countries in the company of a scholar is a big perk. Our clients are really fabulous people, and getting to know the ones who travel with us over and over again means always making new friends. Negatives: really there aren't any except when we are very, very busy it can be stressful. It sometimes seems that we are never able to complete our ever-expanding list of things to do.

Q. What is the future employment outlook for the field?

A. Not good right now due to the economy. But it also depends upon what facet of this industry is desired. I started a company, and I doubt that there are many people who would be interested in doing this. However, to be a tour manager or travel agent, with some training and appropriate skills, jobs are possible.

With experience in all aspects of developing, selling, and leading tours, the ambitious travel specialist might be able to own his or her own company.

WHAT ARE THE SALARY RANGES?

There is very little information available on what adventure-travel specialists earn. Those who work in the field may find that they have peak and slack times of the year that correspond to destination weather conditions or vacation and travel seasons.

Travel guides in all specialties earned salaries that ranged from less than $17,220 to $56,340 or more in 2008, according to the U.S. Department of Labor. The median annual salary was $30,360.

Experienced guides with managerial responsibilities can earn up to $65,000 per year, including tips. Owners of adventure travel businesses can earn $75,000 or more annually.

Travel specialists who work strictly in an office environment may have earnings close to those of travel agents. According to the U.S. Department of Labor, travel agents had median yearly incomes of $30,570 in 2008. The lowest paid 10 percent made less than $18,770 per year, while the highest paid 10 percent earned more than $47,860.

Adventure-travel specialists who work in the field generally receive free meals and accommodations while on tour, and often receive a set amount of money per day to cover other expenses. Major tour packagers and outfitters may offer their employees a fringe benefits package, including sick pay, health insurance, and pension plans.

WHAT IS THE JOB OUTLOOK?

One-half all U.S. traveling adults, or about 98 million people, have taken an adventure trip in their lifetime, according to the *Adventure Travel Report*. This indicates that the market for adventure travel is quite large, and is likely growing despite recent setbacks in the U.S. economy.

Many trends in today's society indicate that this growth is likely to continue. One reason is that the public's awareness and interest in physical health is growing; this leads more and more people to pursue physical activities as a form of recreation. Another reason is that as more people realize that a healthy environment means a better quality of life, there is an increased interest in wildlife and wilderness issues. Adventure travel often encompasses both physical activity and education on and preservation of natural areas, so it is a natural choice for many travelers.

Despite the general growth in the field, however, it should be noted that jobs as tour guides may not be easy to come by. Compared to the rest of the travel market, the adventure segment is still fairly small. Perhaps more significantly, tour guide positions are considered very desirable. Job openings for fieldwork in adventure travel are somewhat limited and highly sought after.

Karen Cleary believes that the employment outlook for the adventure travel industry is "moderately positive. The travel industry has faced numerous challenges; every year or two it is something new, with the [2008] economic crisis an obvious issue for some companies. Firsthand knowledge of destinations is valuable in this field, and well-traveled individuals who are also business savvy can settle into a very nice niche."

Bed and Breakfast Owners

SUMMARY

Definition
Bed and breakfast owners, either single-handedly or with the help of spouse and family, provide guests with a comfortable, home-like environment.

Alternative Job Titles
Innkeepers

Salary Range
$7,000 to $75,000 to $168,000+

Educational Requirements
High school diploma

Certification or Licensing
Required by certain states

Employment Outlook
About as fast as the average

High School Subjects
Business
Family and consumer science
Mathematics

Personal Interests
Business management
Cooking

There's more to running a bed and breakfast than just checking in customers, making breakfast, and cleaning rooms. And if you don't believe it, just talk to innkeeper Martha Hall, who says that a day in the life of an innkeeper can be both challenging and rewarding. Martha is the owner of The Arcadian Inn Bed & Breakfast (http://www.arcadianinn.com) in Edmond, Oklahoma. The inn has been named the best bed and breakfast in Oklahoma nine times by the readers of *The Oklahoman* (the largest newspaper in the state). Below, Martha breaks down a typical day in her life at The Arcadian Inn.

Morning

6 A.M.: Wake up. Pick up cordless phone and keep it with me for the rest of the day. Do not be without your phone.

6:30 A.M.: Breakfast preparation. We have help, but you need to be awake at least an hour and a half before guests to turn on lights, have coffee ready, and get the "smell" of breakfast going throughout the house.

8 to 9:30 A.M. (unless a guest needs to leave earlier, then we serve early): Serve breakfast. We prepare each room service order fresh and deliver it exactly at the pre-arranged time.

We knock on the door, greet our guest with a cheery good morning and smile, set up the table, light candles, announce breakfast, announce check-out procedure, and slip out.

9:30 to 10 A.M.: Clean up kitchen and do any prep for tomorrow's breakfast

10 A.M. to 12 noon: This is check-out time for guests. We give a check-out thank you with a picture of the room, then move to the office to check for online reservations, answer emails, prepare paperwork for check-out, and take care of gift shop needs.

12 noon: Last guest out. We go strip rooms, get washers going, assign housekeepers, or clean rooms ourselves.

Afternoon

12 noon to 4 P.M.:

- Keep tabs on computer for online reservations and guest inquiries
- Keep laundry going
- Check rooms after housekeepers
- Answer phones and doorbells
- Give tours
- Do routine maintenance and lawn care
- Do any redecorating or seasonal decorating
- Prepare inn for guest check-ins, have check-in paperwork ready, names on welcome board, personalized welcome letters and candy on beds, music playing in all rooms, rooms lit to welcome guests, and refreshments set out in dining room

4 P.M. to 8 P.M.: Check in guests with a happy welcome, visit with them, show them around the inn and to their room; ask about anything you can get for them, then do it

Evenings

6 P.M. to 10 P.M.:

- Wait on guests, take ice to guests, make dinner reservations, visit on porch with them
- Marketing, study ads, and make a plan for advertising
- Update Web site and Web directories
- Work on projects such as bed and breakfast association work, community work, etc.
- Do bookkeeping, pay bills, adjust budget, reconcile accounts

10 P.M.: Put the house "to bed," adjust heat/air conditioning, turn down lights, lock doors, put leftover refreshments away

10:30 to 11:30 P.M.: Head out to do grocery and supply shopping while husband takes phone duty

Midnight: Kiss hubby goodnight

WHAT DOES A BED AND BREAKFAST OWNER DO?

A bed and breakfast is an inn, or small hotel, of about four to 20 rooms. The Professional Association of Innkeepers International (PAII), a professional association for the owners of bed and breakfasts and country inns, classifies the different kinds of bed and breakfasts.

A host home is considered a very small business with only a few rooms for rent. Because of its small size, the owner of a host home may not be required by law to license the business or to have government inspections. Without advertising or signs, guests are referred to these homes primarily through reservation service organizations. A bed and breakfast and bed and breakfast inn are classified as having four or more rooms. They adhere to license, inspection, and zoning requirements and promote their businesses through brochures, print ads, and signs. A country inn is considered a bit larger, with 10 or more rooms, and it may serve one meal in addition to breakfast. There are approximately 20,000 bed and breakfasts in the country. Though a bed and breakfast may be located in the very heart of a large city, most are located in small towns, the country, and along oceans, lakes, or rivers. Most of the bed and breakfasts across the country are housed in historical structures: the Victorian houses of Cape May, New Jersey; Brooklyn brownstones; a house in Illinois designed by Frank Lloyd Wright. And many are furnished with antiques.

Imagine yourself living in a beautiful, restored historical house among antiques and vacationers from all around the world. And you don't have to leave to go to work. Though it sounds like an ideal environment, and it may not seem like you're at work, *bed and breakfast owners* must perform many responsibilities to keep their property nice and pleasant. Their chores are mostly domestic ones, keeping them close to the house with cooking, cleaning, gardening, and laundering. This makes for a very comfortable work environment over which they have a great deal of control. Though bed and breakfast owners work in their own homes, they must sacrifice much of their privacy to operate their business. They must be available to their guests at all times to ensure that their stay is comfortable. However, even the most successful bed and breakfast isn't always full to capacity, and many are only open on weekends; this may result in a few long workdays, then a few days of downtime. But to keep their business afloat, bed and breakfast owners will need to welcome as many guests as they can handle.

Among all the daily tasks, bed and breakfast owners also actively interact with guests to make sure they're enjoying their stay and provide information about tours, museums, restaurants, theaters, and recreational areas. It is such close attention to detail that makes a bed and breakfast successful. The guests of bed and breakfasts are looking for more personal attention and warmer hospitality than they'd receive from a large hotel chain.

Bed and breakfast owners handle a variety of business-related concerns such as answering email messages, calling prospective guests, and taking reservations. They must also do accounting work, prepare advertising brochures for the mail, and keep their business's Web site updated.

Though the owners of bed and breakfasts are giving up much of their privacy by allowing guests to stay in the rooms of their own homes, they do have their houses to themselves from time to time. Some bed and breakfasts are only open during peak tourist season, and some are only open on weekends. And even those open year-round may often be without guests. For some owners, inconsistency in the business is not a problem; many bed and breakfasts are owned by couples and serve as a second income. While one person works at another job, the other tends to the needs of the bed and breakfast.

WHAT IS IT LIKE TO BE A BED AND BREAKFAST OWNER?

Scott Nickel (along with his wife Truanna) is the owner and operator of Brickyard Barn Inn Catering & Event Planning in Topeka, Kansas. The Brickyard Barn Inn is a converted dairy barn that was built in 1927. (Visit http://www.brickyardbarninn.com to learn more about the inn.) "We are the seventh owners and have been here almost 11 years," says Scott. "We have continued the B & B concept, but do a great deal of food and events here, including weddings, rehearsal dinners, showers, corporate events, retreats, church functions, etc. We are caterers away from the inn as well as doing the food at the inn."

Scott says that there are no typical days for innkeepers. "Usually my day depends upon whether or when the guests are coming," he explains. "Assuming guests are coming in, I will ensure that the inn is correctly cleaned and the rooms are ready for guests. This is important due to the small staff we have (many times they do the 'dirty work' for me, but I have to do quality control), and because of the events that we do. After I have done my checks, there is marketing to attend to (checking with Internet sites, print ads, etc., and developing new materials), phone calls to catch up on and return, paperwork to keep up with, taxes to file and pay, yard work during various seasons, cleaning pools and hot tubs, fixing whatever is broken, meeting with appointments, meeting with the occasional person who doesn't have an appointment, meeting with salesmen, etc. Making time for family is also critical. My main duties involve seeing to the guests' comfort and ensuring that the inn presents a positive appearance to guests and the public. My secondary duties include marketing, employee paperwork, tax filings, cash flow, and ensuring that I have a life!"

Paige Olson is the proprietor of Kinni Creek Lodge & Outfitters in River Falls, Wisconsin. (Visit http://www.kinnicreek.com to learn more about the lodge.) "I established the business in 2000," she says. "I started out with three guest rooms and six canoe rentals. Then over the years I expanded the options to include the B & B as a total vacation or cabin rental, too. I now have 40 kayak rentals. Over the years as fly fishermen came to stay they wanted guides, the guides wanted a fly shop, and I opened a fly shop for them.

"I offer guided fly-fishing trips, kayak tours, and longer paddling overnight camping adventures," she continues. "The fly fishing and kayaking are done mostly on the Kinnickinnic River, right

Pros and Cons

Scott Nickel, owner of the Brickyard Barn Inn in Topeka, Kansas, details what he likes most and least about his job.

Pros

- Self-employment—building a business, making your own hours, setting your own goals, etc.—is a great feeling.

- I really like the atmosphere of self-employment. I had always hated being part of office politics, petty "rules," office infighting, personality disputes, etc.

- I get to meet all types of people. I enjoy that a great deal.

- Each day is different and unique.

Cons

- All of the pros are two-edged swords. Self-employment is also tough. It requires long hours and self-motivation.

- Sometimes I hate the atmosphere: sometimes being the boss makes me deal with staff issues like office politics, infighting, personalities, etc.

- I get to meet all types of people, some of whom aren't pleasant.

- Each day is different and unique. Sometimes that isn't a good thing. I am a list maker, and occasionally the day-to-day emergencies supersede my list of goals.

- Dealing with the "customer is always right" concept.

- Dealing with the public that doesn't understand that the innkeeper won't have a staff to greet them if they happen to decide to be late. Usually that happens when the innkeeper has theatre tickets, a family event, etc.

- Sometimes when the money doesn't come in as fast as it goes out, owners get nervous.

outside. I offer two sections of river for kayak trips and access to several coldwater spring creeks in Pierce County for fly fishing. The overnight adventure trips are further north on the Brule River and in the Boundary Waters Canoe Area Wilderness of Minnesota."

Paige says that one of the most rewarding aspects of her job is "watching families and friends come together to share some leisure time in the outdoors and create memories that will last a lifetime. Our society has become so mechanized and techno-minded that people forget to

get outside and take some time to move their bodies in the woods, the water, and among friends and family. It is essential for good living."

Mike Hohner (along with his wife Gayle) has owned The Hillcrest Inn Bed and Breakfast in Burlington, Wisconsin, for more than 12 years. It is centrally located to three major cities (Chicago, Milwaukee, and Madison), and is only a short drive from Lake Geneva, Wisconsin, the third-largest tourist destination in the state. (Visit http://www.thehillcrest inn.com to learn more about the inn.)

"After successful careers in education," says Mike, "we were looking to change careers. We wished to do something different and something together. Purchasing a B & B and running it has fulfilled both these desires.

"Bed and breakfast owners," he continues, "need to display a wide variety of talents in carrying out their duties: cooking/food prep, cleaning, organizing, bookkeeping, marketing, decorating, greeting, concierging, lawn care/snow removal, communicating, repairing, networking, selling, etc. Each is of equal importance because you never know what the next phone call or guest might bring. The interpersonal relationships have been the most fulfilling aspect of being an innkeeper. Our least fulfilling is the amount of time we spend waiting for guests to arrive; whether it be for check-in, breakfast, a scheduled event, or check-out."

Mike says that one of the most rewarding aspects of being a bed and breakfast owner are the contacts he and his wife have made with people of varying ethnicities, cultures, and nationalities. "This has assisted us in understanding and accepting the world we live in," he says. "Also, learning how our lives affect and are affected by a global network. People the world over are good and have treated us and our business with respect. We hope that we have done the same in return."

DO I HAVE WHAT IT TAKES TO BE A BED AND BREAKFAST OWNER?

You will need a diverse set of skills to work as a bed and breakfast owner. You must have a mind for business, but you have to be comfortable interacting with others. You must be creative in the way you maintain the house, paying attention to décor and gardening, but you should also have practical skills in plumbing and other household repair (or you should at least be capable of diagnosing any need for repair). A knowledge of the electrical wiring of your house and the phone lines is valuable. "You have to be an accountant, computer expert, electrician, plumber, carpenter, marketing guru, chef, caterer, boss, and psychologist," says Scott Nickel. "You have to be able to do a little of a lot of things as you can't afford to hire many things out." You'll also need an ability to cook well for groups both large and small.

Bed and breakfast owners should enjoy meeting new people. You'll be expected to be a gracious host to all your guests. "You must have a love of service and people of all types," says Scott. "Nothing else matters. You must not mind getting up one and a half hours earlier than your guests, cleaning a toilet after the guests have checked out on a Saturday or Sunday afternoon, meeting a guest three hours after they said they would arrive when you have theatre tickets, dealing with a guest that spilled wax on your carpet, etc. If you walk into innkeeping with blinders or rose-colored glasses, you are in for a rude awakening. The average lifetime for owning a B & B is three to five years here in Kansas. I have been an owner almost 11 years."

Bed and breakfast owners also need to be able to maintain rules and regulations; guests of bed and breakfasts expect a quiet

To Be a Successful Bed and Breakfast Owner, You Should...

- have good business skills
- have a friendly personality
- be highly organized
- be able to cook and bake
- enjoy meeting new people
- not mind working odd hours to keep your guests happy
- have patience—sometimes guests do not arrive on schedule or make challenging requests
- have good marketing skills

environment, and smoking and drinking is often prohibited.

If turning your home into a bed and breakfast, you should learn about city planning and zoning restrictions, as well as inspection programs. Computer skills will help you to better organize reservations, registration histories, and tax records. You should have some knowledge of marketing in order to promote your business via ads, brochures, and on the Internet.

HOW DO I BECOME A BED AND BREAKFAST OWNER?
Education

"I fell in love with the outdoors and the potential to play every day when I was three years old on a rainy, cold camping trip with my family," recalls Paige Olson. "In high school, as I was preparing to choose a college, I learned that I could in fact major in outdoor recreation. I started my first two years at the University of Wisconsin—Stevens Point in youth programming and camp management, then transferred to Colorado State University to complete a BS degree in natural resources recreation and tourism. Then I went onto Prescott College in Arizona for my master's in adventure education and wilderness leadership."

High School

Because you'll essentially be maintaining a home as a bed and breakfast owner, you should take home economics courses. These courses can prepare you for the requirements of shopping and cooking for a group of people, as well as budgeting household finances. But a bed and breakfast is also a business, so you need to further develop those budgeting skills in a business fundamentals class, accounting, and math. A shop class, or some other hands-on workshop, can be very valuable to you; take a class that will teach you about electrical wiring, woodworking, and other elements of home repair.

Postsecondary Training

As a bed and breakfast owner, you're in business for yourself, so there are no educational requirements for success. Also, no specific degree program will better prepare you than any other. A degree in history or art may be as valuable as a degree in business management. Before taking over a bed and breakfast, though, you may consider

Bed and Breakfast Owner Profile: Jim Holder

Jim and Brenda Holder are the owners of the Historic Hayes House Bed and Breakfast in Muskogee, Oklahoma. (Visit http://historichayeshouse.com to learn more about the inn.) Jim discussed the field with the editors of *What Can I Do Now? Travel and Tourism*.

Q. Can you please tell us a little about yourself and the Historic Hayes House?

A. My wife Brenda and I bought Hayes House in 1996 as our home to retire to. I had been in the ministry for 30 years and lived in a parsonage for all those years. We had no home of our own because of living in the parsonage. In 1995 I had open-heart surgery and decided it was time to plan for my future. We found the house in good shape but boarded up and vacant for 10 years, so it needed a lot of TLC.

Hayes House was originally built by Oscar Hayes, who desired to be the first governor of Oklahoma in the year of statehood, 1907. He would tell people in his speeches that when he won the race he would give his inaugural speech from the balcony of his new governor's mansion in Muskogee, Oklahoma. He lost the race.

Q. What did you need to do in terms of rehabbing/renovating your property after its purchase?

A. The house was about 90 years old, so we decided to gut the house and replace all the wiring, plumbing, and gas lines, as well as take all the plaster and lath off the walls, put in insulation, and then hang sheetrock on the walls. We did all the work ourselves. My wife was the decorator (she makes drapes and is an excellent seamstress) and chose all the colors and designs that I had to make and paint. What I did not know how to do, I learned. I developed this theory: If you are not willing to do a thing at least three times, hire it done. Anytime you do something for the first time the "right way," you don't do it right and you know it. So you try it again, the third time. If you are not willing to go through this learning experience, let the professionals make their money. This process is rewarding and highly profitable. Now you know how to do something you never knew how to do and you still have money to do other things.

I must say that there are some downsides to any project like this. You have heard of the "money pit." Get a structural engineer's report before you ever buy any old house. We were very fortunate; however, we still ran into some things that we were not ready for. So will you!

Q. Why did you decide to enter this career?

A. When we began the remodeling process we intended to take 20 years to complete the process. We were funding the remodeling out of pocket for the most part, a little at a time. After six years, an investor made an offer to buy our church property, which included two parsonages, a day care, and the church. The church decided to sell and move the church. They sold my parsonage as

(continued on next page)

(continued from previous page)

well. The board asked me if I wanted to live in my home or in another parsonage. I said I wanted to live in my home, but it was not ready to live in yet. I had a year to make it livable. So we borrowed money and accelerated the remodeling. With the borrowing of more money, my mortgage began to grow. Since I had not been used to making a house payment, and I didn't feel comfortable asking the board to give me more money for *my* house, we began to consider ways to make up the deficit. A friend in town owned a B & B and suggested we start a B & B to make up the mortgage. We did and have loved the venture.

Q. Take us through a day in your work life. What are your main and secondary job duties?

A. I get up at 6:30 A.M. every morning because that is when my body clock goes off. I prepare the breakfasts because my wife decided that because her body clock does not go off that early, she would teach me how to cook. Brenda makes sure the rooms are clean, but I do the kitchen duties. After breakfast is done and the dishes are cleaned and rooms are cleaned and laundry is processed, it is usually 11 A.M.; we run any errands we need done and wait for guests to arrive after 3 P.M. Evenings are spent ironing sheets and napkins while watching our favorite shows on TV. Not hard work, but consistent work.

Q. What are the most important personal and professional qualities for bed and breakfast owners?

A. If you don't love people, don't get into this business. People can really get on your nerves. Also, you must be a self-starter. You cannot wait for someone to tell you that something must be changed or fixed. Patrons will tell you when they are not happy. Maintenance of the facility is a must! Being handy with fixing things is a plus so that you don't spend all your profits on repair people.

Q. What are some of the pros and cons of work in this field?

A. The pros have to do with the element that people come and pay you to see the work you have put into the house that you love. Everyone loves to brag about all that they have done on their home, but this profession is so neat because of the fact that you get paid to talk about your accomplishments.

The cons would be that you have someone in your house all the time. When family wants to visit they have to call ahead of time and "book" a time to come to see you. Beyond these things, I love the B & B business.

enrolling in a hotel management or small business program at your local community college. Such programs can educate you in the practical aspects of running a bed and breakfast, from finances and loans to health and licensing regulations.

Opportunities for part-time jobs with a bed and breakfast are few and far between. Bed and breakfast owners can usually use extra help during busy seasons, but they can't always afford to hire a staff. Some, however, do enough business that they can hire a house-keeper or a secretary, or they may have an extra room to provide for an apprentice willing to help with the business. "Employment in the B & B industry is usually quite limited and stable," says Mike Hohner. "Most owners do the tasks themselves or hire out a few of the less pleasant tasks. Of the B & B owners we are familiar with, any hired staff is usually long-term, which presents few openings or opportunities for aspiring B & B workers. To become B & B owners, though, the opportunities are endless."

Certification or Licensing

Though bed and breakfast owners aren't generally certified or licensed as individuals, they do license their businesses and seek accreditation for their inns from professional organizations such as the PAII. With accreditation, the business can receive referrals from the associations and can be included in their directories. A house with only a room or two for rent may not be subject to any licensing requirements, but most bed and breakfasts are state regulated. A bed and breakfast owner must follow zoning regulations, maintain a small business license, pass health inspections, and carry sufficient liability insurance.

Internships and Volunteerships

If you enroll in a formal hospitality management program, you will likely be required to participate in an internship at a hotel or bed and breakfast. You might work as a desk clerk, restaurant manager assistant, or in another department. Participating in an internship is an excellent way to learn more about the field. You can also volunteer at a hotel or bed and breakfast to get an idea of the jobs that are available.

WHO WILL HIRE ME?

Innkeepers are self-employed. The charm of bed and breakfasts is that they are owned and operated by individuals, or individual families, who live on the premises. Though bed and breakfast "chains" may be a thing of the future, they are not expected to greatly affect the business of the traditional "mom and pop" operations.

Most bed and breakfasts are located in rural areas and small towns where there are no large hotels. Though the number of inns in cities is increasing, only 12 percent of the inns in the United States are located in urban areas. According to PAII, the majority of inns (54 percent) are in small resort villages. Twenty-nine percent of the inns are in rural areas.

An innkeeper's income is derived from room rental and fees for any "extras" such as additional meals and transportation. An inn's guests are often from outside of the local area, but an inn may also cater to many area residents. Most guests are screened by reservation service organizations or travel associations; this helps to protect both the guest and the owner. Bed and breakfasts must pass certain approval requirements, and guests must prove to be reliable, paying customers.

All the bed and breakfast owners you speak to will probably have different stories about how they came to own their businesses. Some convert their own homes into inns; others buy fully established businesses, complete with client lists, marketing plans, and furnishings. Others inherit their bed and breakfasts from family members. And still others lease a house from another owner. Usually, bed and breakfast ownership requires a large investment, both in time and money. Before starting your business, you must do a great deal of research. Make sure the local market can support an additional bed and breakfast and that your house and grounds will offer a unique and attractive alternative to the other lodging in the area. Research how much you can expect to make the first few years, and how much you can afford to lose.

Established bed and breakfasts for sale are advertised nationally, and by innkeeper associations. Prices range from under $100,000 to more than $1 million. An established business is often completely restored and includes antique furniture and fixtures, as well as necessary equipment.

WHERE CAN I GO FROM HERE?

Successful bed and breakfast owners may open additional properties or expand an existing one. In many cases, a married

---

Rewarding Moments

Martha Hall, owner of The Arcadian Inn Bed & Breakfast, details some of her most rewarding experiences in nearly two decades as an innkeeper:

- We've met lots of dignitaries, but the most rewarding is serving guests who really appreciate what we do, who think that we are the most wonderful thing that ever happened to them.

- The bride and groom whose wedding across the street was held during a terrible ice storm and we went to their wedding to give them some guests and hosted their wedding night.

- Being home away from home to many corporate guests, one whose husband called me to find out what his wife's schedule was that day is rewarding.

- When our frequent guest, an Oklahoma City police officer, was shot by a suspect, I was so honored that his wife brought him to our inn to relax and get away from all the stress and publicity to recover.

- I enjoyed helping a speaker from Disney Institute mend his tie; he then used this as an example of "bumping the light"— or extra-thoughtful customer service.

- Helping other people become innkeepers, teaching, mentoring, and even building their inns with them is very fulfilling.

- I like being an important part of the community, serving as chairman of the board of our convention and visitor's bureau, and making presentations to city council on behalf of the board.

bed and breakfast owner may continue to work full time outside of the home while his or her spouse sees to the daily concerns of the inn. But once a business is well established with a steady clientele, both spouses may be able to commit full time to the bed and breakfast.

WHAT ARE THE SALARY RANGES?

Large, well-established bed and breakfasts can bring in tens of thousands of dollars every year, but most owners of average-sized inns must make do with much less. A survey by the PAII provides a variety of income figures. A beginning bed and breakfast has an annual net operating income of $25,000, while one seven years or older has an average income of over $73,000. A small bed and breakfast with four rooms or fewer for rent has an annual net income of about $7,000; an inn of five to eight rooms has an income of $35,000; nine to 12 rooms, $80,000. An inn with 13 to 20 rooms has a net operating income of over $168,000.

Bed and breakfasts in the western part of the United States make more money than those in other parts of the country. An average net income of $68,000 per year is figured for inns in the West, followed by $58,000 for those in the Northeast, $38,000 in the Southeast, and $33,000 in the Midwest. According to PAII, bed and breakfasts charge from $38 to $595 per day, depending on size of the room and whether it has a private bath, fireplace, and other amenities.

Bed and breakfast owners must provide their own benefits, such as health and life insurance and a savings and pension program.

WHAT IS THE JOB OUTLOOK?

Some bed and breakfasts have been in business for decades, but it's only been in the last 20 years that inns have become popular vacation spots. PAII estimates the number of inns in the country to be approximately 20,000, up from a measly 5,000 in 1980. Tourists are seeking out inns as inexpensive and charming alternatives to the rising cost and sterile, cookie-cutter design of hotels and motels. People are even centering their vacation plans on bed and breakfasts, booking trips to historical towns for restful departures from cities. As long as bed and breakfasts can keep their rates lower than hotel chains, they are likely to flourish.

Recognizing the appeal of bed and breakfasts, some hotel chains are considering plans to capitalize on the trend with "inn-style" lodging. Smaller hotels composed of larger, suite-style rooms with more personalized service may threaten the business of some bed and breakfasts. But the charm and historic significance of an old house can't easily be reproduced, so bed and breakfasts are expected to maintain their niche in the tourism industry.

The Americans with Disabilities Act (ADA) will also have some effect on the future of bed and breakfasts. Inns with more than six rooms are required to comply with the ADA, making their rooms and grounds handicapped accessible.

When purchasing a property for the purpose of a bed and breakfast, buyers must take into consideration the expense and impact of making such additions and changes. Though some businesses may have trouble complying, those that can will open up an area of tourism previously unavailable to people with disabilities.

Cruise Ship Workers

SUMMARY

Definition
Cruise ship workers provide services to passengers on cruise ships.

Alternative Job Titles
(The following job titles are just a sampling of the opportunities that are available in the cruise industry.)
Bartenders
Captains
Chefs
Chief pursers
Cruise directors
Physicians

Salary Range
Varies by specialty

Educational Requirements
Varies by occupation

Certification or Licensing
Required for certain positions

Employment Outlook
About as fast as the average

High School Subjects
Business
Foreign language
Geography

Personal Interests
Boating
Entertaining/performing
Helping people: personal service
Travel

"My favorite part of the job—without a doubt—is performing on stage," says John Heald, the senior cruise director for Carnival Cruise Lines. "Carnival cruise directors of old used to have a specific act. This meant they had to be singers, comedians, magicians, etc. I was the first cruise director who was . . . well . . . talentless. However, I do enjoy getting guests on stage and making them the stars. Yep, being on stage is the best part of the job."

WHAT DOES A CRUISE SHIP WORKER DO?

Many modern cruise ships are similar to floating resorts. They offer fine accommodations, gourmet dining, and every possible activity and form of entertainment. It takes a staff of hundreds, and sometimes thousands, to ensure the smooth operation of a cruise ship and the comfort of all passengers. All employees, regardless of their rank, are expected to participate in routine

To Be a Successful Cruise Ship Worker, You Should...

- have excellent communication skills
- be outgoing
- be a hard worker
- enjoy working with people
- be well-groomed
- act professionally at all times
- be willing to work away from home for months at a time

lifesaving and safety drills. Crew organization is divided into six different departments (smaller liners may not have as many divisions of organization); the *Captain,* or the *Master of the ship,* oversees the entire crew.

Deck. This department is responsible for the navigation of the ship and oversees the maintenance of the hull and deck.

Engine. This staff operates and maintains machinery. Together, deck and engine staffs include officers, carpenters, seamen, maintenance workers, electricians, engineers, repairmen, plumbers, and incinerator operators.

Radio department. Videographers are responsible for the maintenance and operation of the ship's broadcast booth, including radio and news telecasts.

Medical department. *Physicians* treat passengers whose maladies range from seasickness to more serious health problems. *Nurses* assist the doctors and provide first aid.

Steward. This department, one of the largest on board, is concerned with the comfort of all passengers. The food staff includes specially trained *chefs* who prepare meals ranging from gourmet dinners to more casual poolside fare. The *wait staff* serves guests in the formal dining room and provides room service. *Wine stewards* help passengers with wine choices and are responsible for maintaining proper inventories aboard the ship. *Bartenders* mix and serve drinks at many stations throughout the ship. From simple blocks of ice, *sculptors* create works of art that are used to decorate dining room buffets. The housekeeping staff is composed of *executive housekeepers* and *room attendants* who keep cabins and staterooms orderly, supply towels and sheets, and maintain public areas throughout the ship.

Pursers. This large department is responsible for guest relations and services. The *chief purser,* much like a hotel's general manager, is the head of this department and is the main contact for passengers regarding the ship's policies and procedures. *Assistant pursers,* considered junior officers, assist the chief with various duties, such as providing guest services, ship information, monetary exchange, postage, safety deposit boxes, and other duties usually associated with the front desk department of a hotel. The *cruise director,* also known as *entertainment director*, heads the cruise staff and plans daily activities and entertainment. The *youth staff director* plans activities and games specifically designed for children. Ships with a casino on board employ casino workers, including *game dealers, cashiers, keno runners*, and *slot machine attendants*. *Sound and lighting technicians* are needed to provide music and stage lighting for the many entertainment venues found on board. Many *entertainers* are hired to sing, dance, and perform comedy skits and musical revues. *Dance instructors* teach dance classes ranging from ballroom to country. *Fitness instructors* teach aerobics and other exercise classes. Also, many employees are hired to work in duty-free shops and souvenir stores, beauty parlors, spas, health clubs, and libraries.

Other occupations in the cruise ship industry include clerical workers, human resources workers, computer specialists, and security workers.

Workers in the cruise line industry do not have a lot of free time. They are there to work, not enjoy the amenities that are available to guests. Most cruise ship workers work long hours—eight- to 14-hour days, seven days a week are not uncommon. Many employees spend a number of weeks, usually five or more, working at sea, followed by an extended leave ashore.

Being a people person is important in this industry. Cruise ship workers not only are expected to work well with their coworkers; they have to live with them, too. Accommodations for the crew are especially tight; usually two to four employees are assigned to a room. The crew has dining areas and lounges separate from the passengers, yet total privacy is rare on a cruise ship. Crew members usually have little access to public areas on their free time. However, when the ship docks at port, crew members on leave are allowed to disembark and go shoreside.

WHAT IS IT LIKE TO BE A CRUISE SHIP WORKER?

John Heald is the senior cruise director for Carnival Cruise Lines, the largest cruise line in the world (based on the number of passengers carried annually). "I started with Carnival Cruise Lines back in 1987," he recalls. "I was at the time working for Lloyds of London and, although it was a well-paid job with great prospects, I knew in my heart that I was meant to do something else. I then saw an advertisement in a magazine for people who were required to work as bar staff onboard a cruise liner. On a whim, I went for an interview and three weeks later, I was flying from London to Miami. I was the worst bartender in the world and this was quickly recognized by Carnival, which told me I was dreadful because I spent more time entertaining the passengers rather than serving drinks. So, they told me that instead of being in the bar department I would be working as a social host, who is part of the entertainment department. They handed me a microphone, something I had no experience with, and two years later I became a cruise director and here I am some 20 years later still here and still enjoying the best job in the world.

A Rewarding Experience

John Heald details one of his most gratifying experiences as a senior cruise ship director:

I have so many stories to tell but none sum up the rewards of this career better than this. One day while the ship was in my home port, my mail arrived and among the letters was a box. Inside the box was a beautiful gold plaque. It simply said "To John Heald, Cruise Director, who reminded us how to laugh." With the plaque was a letter from a guest who had sailed a month or so before. I had not met them or spoken with them. However, they had sent me this plaque because they had laughed for the first time in three years, and the first time since their seven-year-old daughter was killed by a drunk driver. That's why this is the best job in the world, and I wish everyone reading this the very best of luck in whatever path their lives take them.

"Describing a typical day as a cruise director," John continues, "is about as difficult as understanding Klingon. That's because there really is no such thing. Yes, the shows you host and the staff you manage may be the same each cruise, but with 3,000 guests on board each week there is always going to be something new to experience. The cruise director is in charge of all the entertainment and activities and information given to guests. A day at sea will start at 8:00 A.M. and involves producing the daily program, meeting with staff to discuss the entertainment and activities for the day, and hosting said activities and shows. I also must be visible and out talking to guests. I manage a department of 70 staff from many different countries and, therefore, the ability to be a dad, big brother, and a boss is very important. The evening consists of more shows and usually finishes by 1:00 A.M., and that does not include the many emails and computer work that is the unseen part of the job. The days are long, but very rewarding."

DO I HAVE WHAT IT TAKES TO BE A CRUISE SHIP WORKER?

You will need to be at least 21 years of age and have a valid U.S. passport to work in this field. If you hold a passport from another country, you will need to obtain a work visa. Check with your country's embassy for details and requirements.

Besides having the proper education, experience, and credentials, employers look for applicants who are outgoing, hardworking, friendly, enjoy working with people, are able to follow instructions, and have excellent communication skills. It is important to make a positive impression with the passengers, so cruise ship workers should always be neatly dressed, properly groomed, and well behaved at all times. Inappropriate contact with passengers is not tolerated.

"It would be easy to say the most important qualities in a cruise director should be that they are 'a people person,'" says John Heald. "For me, though, that is a throwaway line and does not even come close to what a great cruise director should be all about. He or she must love life. You can never have a bad day at work, never ever. You might be mad at the world, or have a very bad headache, but no matter how bad you feel, the moment you walk out of your cabin into a guest area, you must be full of life and happiness. You see, you are never off duty and that may seem easy, but I promise you it is not. So, if you love life, are someone who is 100 percent positive all the time, and can speak clearly and in an entertaining way, then this job may be for you."

HOW DO I BECOME A CRUISE SHIP WORKER?

Education

High School

You will need at least a high school education, or equivalent, to qualify for most entry-level jobs. While in high school, you should concentrate on classes such as geography, sociology, and a foreign language. Fluency in Spanish, French, and Portuguese is highly desirable.

Postsecondary Training

Officer-level positions, or jobs with more responsibility, require college degrees and past work experience. Many employees, especially those on the cruise staff, have an entertainment background. Youth staff members usually have a background in education or recreation. Specialized workers—such as doctors and nurses—must pursue typical educational paths that are required for workers in their fields.

Certification or Licensing

Most entry-level jobs do not require certification. Some technical positions, such as those in the engine room, may require special training. Physicians and nurses must be licensed to practice medicine. Child-care workers should have experience and proper training in child care. Some cruise line employees may belong to the Seafarers International Union.

Internships and Volunteerships

You will most likely not have the opportunity to participate in an internship or volunteer opportunity aboard a cruise ship, but you can obtain similar experience by interning or volunteering at a hotel, amusement park, casino, or other place where large numbers of people seek entertainment. If you are in college, ask one of your professors if he or she is aware of any enrichment opportunities, or contact your college's career services office for more information.

WHO WILL HIRE ME?

Nearly 348,000 people are employed in the U.S. cruise industry, according to Cruise Lines International Association. There are approximately 45 cruise lines with offices in the United States; together, they employ thousands of cruise ship workers. Most employees are contracted to work four or more months at a time. Some major employers include Royal Caribbean International, Carnival Cruise Lines, Princess Cruises, Norwegian Cruise Line, Cunard Line, Holland America, and Disney Cruise Line.

Applicants without college degrees and little shipboard experience are usually assigned to entry-level positions such as wait staff or housekeeping. If you have experience in retail sales, then you may be given a job at the duty-free shop; hospitality experience may land you a position in the purser's office.

John Heald cautions prospective cruise ship workers to be wary of employment scams. "There are lots of people offering advice on the Internet," he says, "and many of them are scams. You send in some money and they supposedly send you a list of contacts. If you feel that a life on board is for you, then please contact the cruise lines directly. Most of the major lines have a careers page on their Web sites, and that is the only way to go. If you decide to enter this career, please prepare yourself for months away from family and friends and know that what you see in the brochure and online (facilities, amenities, etc.) is what the guests

Cruise Ship Worker Spotlight: Ray Rouse

Ray Rouse is the entertainment director for Cunard Line's *Queen Mary 2*. The ocean liner hosts approximately 2,600 guests on a full voyage. Ray has worked in the cruise industry for 35 years. He discussed his career with the editors of *What Can I Do Now? Travel and Tourism*.

Q. Can you detail a typical day on the job?

A. A typical day consists of filming our daily morning show for guests (which is shown on guest television), attending executive meetings with the ship's senior officers, hosting daily activity programs, introducing shows and other theater performances, hosting receptions and cocktail parties with the captain and senior officers, and conducting interviews with celebrity guests and the press.

Q. What do you like most and least about your job?

A. The thing I most like about my job is that there is something different happening every day. I really do not have a "least thing." I feel I have the best job in the world!

Q. What are the most important personal and professional skills for entertainment directors?

A. Leadership and management skills, time management, communication, and interpersonal skills.

Q. What advice would you give to young people who want to enter your profession?

A. Start early at a young age, study hard, and enjoy the company and diversity of the people you work with.

Q. What has been one of your most rewarding experiences during your career?

A. Being involved with the *Queen Mary 2*, the greatest ocean liner in the world, since 2003.

enjoy. There is no doubt that whatever job you apply for, you will work hard, but the rewards are brilliant."

WHERE CAN I GO FROM HERE?

With cruise experience, a cruise staff member can advance to assistant cruise director, and in turn become cruise director. Assistant pursers can be promoted to chief purser. Even people in entry-level positions can be promoted to jobs with more responsibility and, of course, better pay. Bussers can become assistant waiters and then headwaiters. Room stewards can be promoted to housekeeping manager and supervise a team of cleaners or a specific section of the ship.

WHAT ARE THE SALARY RANGES?

There are so many variables that it is hard to gauge the average salary for this industry. First, many employees are hired on a contractual basis—anywhere from four to six months for housekeeping, wait staff, and the concessionaires. The size of the cruise line and the region it sails may also affect wages. According to Cruise Services International, the general salary range is between $1,000 to $1,700 per month. Some employees count on passengers' tips to greatly supplement their income. Restaurant and house staff workers can stand to earn anywhere from $300 to $600 in weekly tips. Workers in professional-level positions, such as captains or physicians, earn higher salaries.

Did You Know?

- The average length of a cruise was 7.2 days in 2008.
- Almost 20 percent of Americans have taken a cruise.
- The average cruise passenger is 46 years of age.

Source: Cruise Lines International Association

Employee benefits include room and board and all meals while on board. Most cruise lines offer emergency health coverage to their employees, regardless of the length of contract. Full-time employees are also offered health insurance, paid sick and holiday time, stock options, and company discounts.

WHAT IS THE JOB OUTLOOK?

The health of the cruise line industry is tied to the state of our nation's economy, as well as the public's perceived level of safety as it relates to terrorism or onboard medical outbreaks (such as the norovirus). Bookings declined significantly in the aftermath of the 9/11 terrorist attacks, and cruise lines upgraded their security measures to ensure the safety of passengers and crew. Cruise line officials also point out that viruses such as the norovirus don't just happen on ships, but in any place where a large number of people are concentrated in a small area.

The cruise line industry is still one of the fastest-growing segments of the travel industry. Approximately 13.2 million people cruised in 2008, according to Cruise Lines International Association (CLIA). Nearly 35 new cruise ships are expected to be added to the North American Fleet from 2008 to 2012, according to CLIA. Ships are getting bigger and more opulent and have become travel destinations in themselves.

With so many mega-ships in operation, qualified cruise ship workers are still in demand. Entry-level positions such as wait staff and housekeeping will be fairly easy to obtain with the proper paperwork and credentials. A college degree and work experience will be necessary for positions with more responsibility. Fluency in French, Spanish, or Portuguese is a plus. A cruise ship will offer workers the opportunity to travel around the world and meet many people of different nationalities and cultures.

Remember, however, that cruise life is not all fun and travel. Cruise ship workers are expected to work long, hard hours, and be away from their home base for weeks at a time. Many people find the schedule exhausting and opt to find employment ashore.

Flight Attendants

SUMMARY

Definition
Flight attendants help airline passengers have a safe and comfortable flight. They prepare the cabin for passenger boarding, help passengers stow luggage and find their seats, and demonstrate the use of emergency equipment. During flight, they serve drinks and prepared snacks or meals, and try to make passengers as comfortable as possible. At the end of a flight, attendants help passengers retrieve their luggage and leave the plane.

Alternative Job Titles
Steward
Stewardess

Salary Range
$20,580 to $35,930 to $65,350+

Educational Requirements
High school diploma

Certification or Licensing
Required

Employment Outlook
About as fast as the average

High School Subjects
Foreign language
Psychology
Speech

Personal Interests
Helping people: personal service
Travel

"The rewards of this career are uncountable," says flight attendant Maria Conner. "In my off hours, I have zip-toured through Costa Rica's jungles, sailed in Hawaii on chartered boats, and toured the Windsor Castle. I have seen volcanoes erupt, felt earthquakes, and met some of the most interesting, exotic, best friends of my life. The day American Airlines pinned my wings on me for the first time was like having the world handed to me on a silver platter. Nothing was going to stop me or keep me from opportunities around the world. Being a flight attendant with American Airlines has given me more opportunities throughout my 17-plus years of service than any other career I would have considered."

WHAT DOES A FLIGHT ATTENDANT DO?

Flight attendants attend to the safety and comfort of airline passengers from the time they board the plane until they leave. Attendants are usually assigned to a base, or hub, which is one of the large cities that their airline flies into and out of. Full-time flight attendants fly approximately 65–90 hours each month, and spend another 50

61

hours on the ground preparing planes for flight, writing reports on completed flights, and waiting for planes that arrive late. Attendants work long days, but overall, they have more days off than employees in standard nine-to-five jobs.

The attendants' responsibilities begin about an hour before the plane takes off, when they report to a briefing session with the rest of the flight crew. At the briefing session, they receive information about weather conditions that may affect the flight and passengers who may have special needs.

Before passengers board the plane, flight attendants check emergency equipment to make sure it is in working order; ensure that the passenger cabins are clean, orderly, and stocked with pillows and blankets; and check the airplane kitchens, or galleys, to make sure that they are supplied with enough food and drinks for the flight.

As passengers board the plane, attendants greet them and help them stow their luggage and coats and find their seats. They may have to give special assistance to passengers who are elderly, disabled, or traveling with small children. Before takeoff, a flight attendant uses a loudspeaker to address the passengers. He or she welcomes them to their flight and gives them any necessary information about delays, weather conditions, or flight times. As required by federal law, flight attendants demonstrate how to use the plane's emergency equipment, and they check to make sure all passenger seat backs are in an upright position and all seat belts are fastened before takeoff.

During the flight, attendants usually serve passengers drinks and either meals or snacks, depending upon the time and the length of the flight. They may also pass out magazines, newspapers, headphones, or pillows and blankets to passengers. They may answer questions, help entertain children, or control unruly passengers. Taking care of passengers who are sick or frightened is another important part of the flight attendant's job.

Although they may never encounter an air emergency, it is very important for flight attendants to know what to do should one occur. In the event of an emergency evacuation, attendants must help passengers leave the plane in a rapid, safe, and orderly fashion. They may have to open emergency doors and inflate emergency slides to allow passengers to evacuate.

Once the plane has landed and taxied to the gate, the flight attendants help the passengers retrieve their luggage from the overhead compartments and exit the plane. On international flights, they may provide customs and airport information and sometimes translate flight information or instructions into a foreign language for passengers. Then they clean up the cabin for the next flight. This includes picking up any trash left behind by the passengers and making sure all the seat belts are straightened. On the last leg of the shift, before the attendants go off duty, they must make an account of any money they have collected for drinks, and drop this off at the airline office. They also speak briefly with the new flight attendants coming on duty to pass on any information they might need to know about the cabin or the galleys.

Attendants may also have some clerical duties. They may collect and account for money made on liquor sales, file incident reports, and fill out forms relating to liquor inventory, lost and found items, or cabin maintenance.

WHAT IS IT LIKE TO BE A FLIGHT ATTENDANT?

Maria Conner has been a flight attendant with American Airlines for more than 17 years. "I wanted to be a flight attendant since I was nine years old," she says. "My family would travel to Puerto Rico once a year for vacation, and my favorite part of the trip was the airplane ride. I would watch the beautiful women dressed in perfectly manicured uniforms walk up and down the aisles. They were always so professional and glamorous. Then I would wonder what they would do when they got to their destinations. Well, now I know."

Maria flies to London, England, once a week on a 777 aircraft—the largest aircraft American Airlines flies. "A typical day for me," Maria explains, "begins when I get to the airport and check in for my flight, which involves signing in and checking to see how many people are in my cabin. I usually fly in first class because I like the more personal one-on-one service. Next, I go to the aircraft and meet my fellow crew members, and then we check the status of our emergency equipment throughout the plane. Each crew member has a position on the aircraft for which

Lingo to Learn

bulkhead The walls on an airplane that divide the cabin into sections.

cabin The passenger compartment of an airplane.

FAA Federal Aviation Administration, the government body that regulates safety standards for aircraft and aviation personnel.

flight deck Also called the cockpit, the area of the plane where the pilot and copilot sit.

galley An airplane kitchen.

hub A city or an airport in which an airline has major operations and many gates.

leg One complete flight, from takeoff to landing.

minimum equipment list (MEL) A list of aircraft equipment that must be in good working order before an aircraft can legally take off with passengers.

pressurized aircraft An aircraft that is kept at a designated atmospheric pressure so that passengers can breathe normally.

turbulence Rough, sometimes violent, atmospheric conditions encountered by airplanes.

they take responsibility. For example, if your position is galley then you set up the kitchen for serving drinks and food. Or if you work the aisle you would set up the videos, newspapers, and headsets." After setting up, Maria and the other flight attendants board the passengers. "Everyone finds their seats, stows their bags, then settles in for the seven-hour trip to London. I try to remember that I have the ability to impact our customers the most since I see them the longest, so I take pride in my service. Once we land and the passengers deplane, the flight crew goes to our hotel and we can spend our layover time as we wish. I like to fly with my friends; we typically shop and try different restaurants.

"When I first started flying at 21," Maria continues, "I couldn't think of any cons in this field of work. However, now that I am older and have a family of my own, spending time away from the family could be my only con. I like to think it is 'mommy time' so it helps me regroup when I come home. The pros of this job are infinite. The cultural experience alone has opened my eyes to the world. I love to learn about different people and their ways of life, so seeing and living it for a couple days a week is my thrill. For example, going to Japan and eating eel bones is quite an experience. I sightsee for a living and get paid for it. Another pro would be exposing my family to many parts of the world. The experiences I have shared with my children are nothing you could learn from a book."

"I started as a stewardess with American Airlines 41 years ago and have become and remain a flight attendant to this day," says Kate Pantorilla. "The two titles to my profession can give you a little insight to the changes over many years. As a stewardess then and flight attendant now my job has steadfastly remained the same, and that is the safety of the traveling public. Yes, there are now better movies, bigger planes, meals/no meals, and better destinations, but my reason for being on the airplane has not varied. Whether you joined the airline 41 years ago or 41 months ago everyone says, 'Oh, I will do this for a short time, travel, and figure out what I want to do with my life.' You become addicted to the time off, the cities, the friends, and the networking that is part of the life.

"The phrase 'typical day' cannot be found in an airline dictionary," she continues. "This coming Saturday I will begin my day with breakfast and get ready for my flight to Shanghai, China; that is when 'typical' stops. With every flight, on paper you go from point A to point B, and that is how the public and sometimes our families see our lives. The minute I sign in (one hour before departure) my world begins by meeting the captain and my fellow flight crew and flight attendants, and the clock is ticking. With boarding the aircraft there are many tasks—from checking emergency equipment (from medical to safety), to catering, preparing entertainment systems, and talking with the agents about the passengers. In one trip I can have a blind passenger, a passenger being deported, passengers traveling because a family member has passed away, or my regular business- men and women that I meet quite often. This can all happen on a three-hour flight to Mexico or a 15-hour flight to India. The flights usually run like 15-minute short films because things are happening every few minutes, and if we are doing our jobs to the best of our abilities as flight attendants you don't notice a thing.

"There are pros and cons to every job, relationship, and location, and if you are lucky in life, like I have been, the pros have it all. I can be sitting on my jump seat before take-off and hear someone say 'Chicago' and I can close my eyes and be taken back to the top floor of the Pick Congress Hotel, where I was invited by a producer of the *Today* show (a passenger) and watching the riots at the Democratic Convention in 1968 unfold on Michigan Avenue. You mention Boston and I remember hearing the Boston Pops play the July Fourth Bicentennial Celebration and the last performance by conductor Arthur Fiedler. Or being in Berlin to get a piece of the wall as it fell. The cons are the early morning get-ups, not being home most holidays the first years, and being on call every three months on reserve, but it has been worth every con to look back at the pros of my career."

DO I HAVE WHAT IT TAKES TO BE A FLIGHT ATTENDANT?

"If this career sounds interesting and exciting," says Maria Conner, "you must have the following qualities. One, the ability to see every situation from beginning to end. You cannot get off the plane or go home [when something goes wrong]. Two, you must be versatile. People and situations change every day, so you need to become a chameleon and change with your environment. Three, you need the ability to stay calm and think on your feet; you may be helping a mom with her baby or saving a man from cardiac arrest. Four, you must be flexible (this is really important). You may have to move to a different city. Or just when you thought your trip was over you have to go to another

> ### Fast Fact
>
> Until the 1940s all of American Airlines' flight attendants were registered nurses. During World War II, however, when nurses were in short supply, American dropped that requirement.

city for another night. And most importantly, you must have the ability to make people feel comfortable and have a streak of adventure."

Being responsible and compassionate are two of the most important characteristics of a good flight attendant. It's also important to be able to assert yourself while still being tactful. You might encounter a person who's had too much to drink. You need to be able to confront that individual in a tactful way in order to calm the situation.

It is important that flight attendants be able to deal effectively with any sort of crisis or emergency situation. Passengers depend on them for help in any situation that arises, whether it is something as minor as airsickness or as major as an emergency evacuation of the plane.

Because they are on their feet for most of their working hours, flight attendants need a certain level of stamina. Because the job does require a certain level of energy and endurance, most airlines require that their flight attendants pass a medical examination and meet certain height and weight standards.

To Be a Successful Flight Attendant, You Should...

- be neat and well groomed
- have physical stamina
- enjoy dealing with people
- be poised, confident, and articulate
- have a warm, outgoing, and compassionate personality
- be controlled, level-headed, and able to respond properly in an emergency

Finally, flight attendants must be willing to accept a slightly different lifestyle than most jobs require. They may be scheduled to work nights, weekends, and on holidays, and they may be away from home for several days at a time. However, because their shifts are longer than the standard eight hours, flight attendants work only three to four days a week.

Most flight attendants belong to one of three labor unions: the Association of Flight Attendants, the Transport Workers Union of America, or the International Brotherhood of Teamsters. In exchange for the weekly or monthly payment of dues, flight attendants who belong to one of these unions receive a package of services designed to improve their working environment. Union services often include collective bargaining for pay and benefits, governmental lobbying, and legal representation.

HOW DO I BECOME A FLIGHT ATTENDANT?
Education

High School

Although many airlines prefer to hire applicants with some college experience, a high school diploma is the minimum educational requirement for this job. Start taking courses that build your communication skills. A flight attendant is the public face of the airline, so airlines want attendants who can speak clearly and professionally. Poor English, grammar, or enunciation may disqualify an applicant. To enhance these skills you should focus on English and speech classes. Classes in a foreign language are also good choices because international airlines usually require their flight attendants to be fluent in at least one second language. "If I were to apply today with an airline I would have at least one language qualification, if not two," says Kate Pantorilla. "I now fly with young people who speak a minimum of three languages, and I do envy them because speaking a foreign language is a talent."

Because so much of the flight attendant's job involves dealing with people, courses in psychology may be helpful. A psychology background may prove especially helpful in dealing with passengers who are frightened or upset. Finally, classes in geography and sociology could help familiarize you with the places to which you may travel on the job.

Aside from choosing classes with an eye toward the future, you can begin to prepare for a career as a flight attendant by finding a summer or part-time job that allows you to work with the public. Jobs in customer service or customer relations are very helpful for aspiring flight attendants.

Postsecondary Training

Because many airlines prefer to hire employees with some college experience, it is advisable to complete a two-year or four-year college degree program. Although there is no specific major that will prepare you for a career as a flight attendant, degrees in psychology, public speaking, sociology, nursing, anthropology, hospitality, police or fire science, travel and tourism, and education are all good choices. A business degree with an emphasis in customer service or public relations is another excellent option. If you are especially interested in international flights, you might consider getting a degree in a foreign language.

Regardless of their previous education, all flight attendants are required by their airlines to complete a four- to seven-week training course. While most large airlines maintain their own schools for flight attendants, some of the smaller airlines may not. These smaller companies often send their flight attendants to schools run by the larger carriers.

During an airline training program, flight attendants learn how to respond in an emergency. They are taught how to administer first aid, how to use the airplane's oxygen system, and how to evacuate the plane in an emergency. Some airlines have a plane on hydraulics that is used to simulate a plane crash. Flight attendants have to go through the steps of sitting there, hearing the crash, and jumping up and performing the evacuation.

Attendants also learn the basics of customer service, grooming requirements, Federal Aviation Administration (FAA) regulations, company operations and schedules, aircraft equipment, and how to fill out flight report forms. Airlines also train attendants in public relations policies such as dealing with customers who have had a death in the family or providing extra help to people with disabilities.

Trainees for international flights are taught how to deal with customs and visa regulations, and what to do in the event of a terrorist attack. Near the end of the training program, attendants go on practice flights in which they perform their duties under supervision. Once they have completed the initial training period, flight attendants must complete 12 to 14 hours of additional training in emergency procedures each year, as mandated by the FAA.

Certification or Licensing

All flight attendants must be certified by the Federal Aviation Administration. To become certified, flight attendants must complete training requirements (such as fire fighting, medical emergency, evacuation, and security procedures) that have been created by the FAA and the Transportation Security Administration. Certification requirements vary for different types of aircraft.

Internships and Volunteerships

If you go to college you will participate in an internship that will give you a general introduction to the travel and tourism industry, but it will probably not be direct experience as a flight attendant (this type

Rewarding Experiences

Flight attendant Kate Pantorilla details some of the most rewarding and moving experiences of her career:

When someone asks, "What has stood out in your career?" I can honestly say that I both laugh and cry [when thinking back]. I met my husband (who is now a retired flight attendant), and we have been married 29 years. We met flying domestic and fell in love in Syracuse, New York, of all places. We flew together many years and shared some wonderful memories.

The second greatest pleasure was being able to fly our troops in and out of the Middle East at the start of the Iraq War. I starting flying during the Vietnam War and I understood the need for our troops to know how important they were and that they had endless support from not only myself, but also American Airlines. Those missions will always have a special place in my heart.

As the world knows so well, American Airlines, along with United, lost two airplanes on September 11, 2001, and if there was ever to be anything "typical" about my job it changed that day. My day started like each flight attendant's day did that morning—with getting ready, kissing family good-bye, and boarding a plane. The only difference is I made it to my hotel room that night, while others did not—and that I live with always.

of experience can only be gained as part of a formal airline training program). You also can volunteer with a travel- or tourism-related organization, but, again, this will not provide firsthand experience in the field. With that said, any experience in the industry will help you get a better idea of work settings and career options.

WHO WILL HIRE ME?

Approximately 97,000 flight attendants are employed in the United States. The best way to start a job search for this position is to compile a list of all major airlines, and contact their personnel departments directly. Although you may be familiar with many of these airlines—such as American, US Airways, United, Delta, Southwest, and Continental—there are many others that are smaller or that operate exclusively in a certain region of the United States. For a complete list of domestic airlines, contact the Air Transport Association of America.

Once you have obtained a complete list, send your resume, along with a cover letter, to each airline you are interested in. Also, many airlines have offices on the premises of major airports. If you are near such an airport, you might consider visiting the airlines' offices to talk with representatives; it may be possible to apply for a position in this manner.

Some of the major airlines have personnel recruiting teams that travel through the country interviewing prospective flight attendants. Airline company offices can provide you with information about these recruitment visits, which are sometimes announced in newspaper advertisements as well.

When a flight attendant is new, he or she is placed on reserve status. Reserve attendants do not have a regular schedule; rather, they must be available on short notice to work extra flights or to fill in for attendants who are sick or on vacation. It may take an attendant between one and 10 years to move out of reserve status, depending upon the number of flight attendants in his or her airline who retire, leave the job, or are promoted.

When the attendant is promoted out of reserve status, he or she may bid on regular assignments for airline bases or flight schedules. Because these assignments are made on the basis of seniority, the longer the attendant has been employed, the more likely it is that he or she will receive an assignment of choice.

Although the vast majority of flight attendants are employed by commercial airlines, there are a small number who work for private companies. Many large corporations, such as IBM, Microsoft, or Boeing, maintain their own aircraft for the purpose of flying their executives from place to place. One or more flight attendants may be present on these jets, depending upon the size. Because most corporations prefer to hire experienced flight attendants, however, finding a job in a private company is unlikely for the beginning attendant.

WHERE CAN I GO FROM HERE?

Flight attendants may advance to lead or first flight attendant (sometimes also called flight purser), supervising flight attendant, instructor, or airline recruitment representative. Others become trainers, interviewers, or recruiters. Some

flight attendants advance by becoming *flight attendant supervisors*. In this position, they spend most of their time in the airline's offices, overseeing other flight attendants and working with the scheduling and payroll departments.

Some flight attendants enter the field after having pursued other careers or, conversely, use their time off to train to enter other fields, such as teaching. "I just flew last week with a crew," says Kate Pantorilla, "and one woman was a lawyer,

Advancement Possibilities

Flight attendant instructors train new flight attendants on aircraft specifications, FAA regulations, emergency procedures, customer service, and first aid. They also conduct refresher training courses for flight attendants already employed.

Flight attendant recruiters may travel around the country to interview prospective applicants for flight attendant positions, represent their airline at career or job fairs, and visit high schools or college campuses to meet with interested students.

Flight attendant supervisors oversee other flight attendants and monitor their performance, serve as a liaison between attendants and airline scheduling and payroll departments, and interview applicants for open positions.

another a grade school teacher, and one a landscaper. This profession does allow you the time to pursue schooling for another job, learn a hobby, or start a family. You set your own limits, not the job."

WHAT ARE THE SALARY RANGES?

Median annual earnings of flight attendants were $35,930 in 2008, according to the U.S. Department of Labor. The middle 50 percent earned between $28,420 and $49,910. Salaries ranged from less than $20,580 for the lowest paid 10 percent to more than $65,350 for the highest paid 10 percent. Wage and work schedule requirements are established by union contract.

Most attendants receive a base pay for a certain number of flight hours each month. They then receive extra pay for overtime and night flights. FAA regulations limit the number of flying hours attendants can work per week, however, so there is a cap on overtime hours. Many airlines pay more for international flights than for domestic ones.

Almost all airlines pay attendants' expenses such as food, ground transportation, and overnight lodging when they are on duty away from home. Airlines require their flight attendants to wear uniforms. Some airlines require new attendants to purchase their own; others supply the uniforms at no cost.

Flight attendants typically receive a standard benefits package, which includes paid sick and vacation time, free or reduced air fare for themselves and family

members, and in some cases, medical and life insurance and a pension plan.

Related Jobs

- airline lounge receptionists
- concierges
- cruise directors
- customer service representatives
- funeral attendants
- hosts/hostesses
- passenger service representatives
- ramp attendants
- reservation and transportation ticket agents
- ship stewards and stewardesses
- social directors of cruise ships
- tour guides
- travel agents
- wait persons

WHAT IS THE JOB OUTLOOK?

Job prospects for this career are expected to be good in the coming several years, with employment of flight attendants predicted to grow about as fast as the average for all careers through 2016, according to the U.S. Department of Labor. The main reason for this growth is an increase in the number of people who are flying. To accommodate the growing number of passengers, airlines are using larger aircraft and scheduling more flights. Since FAA regulations require one flight attendant for every 50 passengers aboard a plane, there should be a heightened need for these workers.

Most job openings will also arise from the need to replace flight attendants who get promoted, leave the field, or switch departments.

Even with the growing need for attendants, however, there is keen competition for these positions. Airlines receive thousands of applications from prospective flight attendants each year, so you will have the best chance of finding a job in this field if you have a college degree and some prior work experience in dealing with the public.

Although the job outlook for flight attendants is expected to be good, it is important to be aware that the airline industry is very sensitive to the overall state of the nation's economy. During economic downturns, people cut back on their leisure travel and many businesses reduce their business travel as well. When the demand for air travel declines, full-time flight attendants may be put on part-time status or laid off, and very few new attendants are hired.

Job opportunities may be strongest at regional and commuter, low-cost, and charter airlines, as well as with companies that have private jets for their executives.

Hotel Concierges

SUMMARY

Definition
The concierge is the hotel's best representative for guest services. They provide services to help make the guest's stay more enjoyable. They offer recommendations regarding the city's best restaurants, shows, museums, and tours. Concierges can also book flights and arrange car or limousine rentals.

Alternative Job Titles
Guest service agents
Guest service representatives

Salary Range
$17,560 to $27,180 to $40,220+

Educational Requirements
High school diploma

Certification or Licensing
Recommended

Employment Outlook
Faster than the average

High School Subjects
Art
English (writing/literature)
Foreign language
History
Speech

Personal Interests
Helping people: emotionally
Helping people: personal service
Music
Psychology
Reading/books
Theater

The boutique hotel Affinia Chicago was undergoing extensive renovations. This proved to be a major challenge for concierge Kathy Smith. "In the summer of 2006," she recalls, "a wedding party decided to stay with us. During this time the hotel consisted of renovated and unrenovated rooms with air conditioning that was either blasting or that did not work. The lobby consisted of a narrow hallway with a cardboard wall on one side and a makeshift plywood counter on the other side. No room service or food was available except the free bottled waters and snacks we had at the counter. Needless to say the wedding party wasn't too happy. They were booked prematurely and thought the renovations would have been completed.

"This scenario was the biggest challenge I had ever faced," she continues. "It took a combination of empathy, anticipation, and the skills of surpassing expectations to make this wedding party happy. We were there handing out water bottles, setting up extra fans in the lobby and changing rooms, arranging transportation, and so on and so forth.

At the end of the day, we were waiting to hear all the complaints from the wedding party, but, instead, we kept hearing compliments. The wedding party was empathizing with us! They couldn't understand how we would be able to make anyone happy while staying at the Affinia amidst the mess.

"However, we did. In fact, our reviews on Tripadvisor.com spoke volumes. Guests over and over again reported the tribulations of staying at a hotel that was undergoing renovations, but they always pointed out that their visit was made enjoyable by the fact that someone on our staff went above and beyond their everyday duties. That moment when a guest smiles and says that you made their trip the best is what the job is about. It's rewarding to make that much of a difference in such as short time with perfect strangers."

WHAT DOES A HOTEL CONCIERGE DO?

Concierges are the most visible and active ambassadors of hotel hospitality. Their basic duty is to provide hotel guests with services to help make their hotel stay as comfortable, enjoyable, and memorable as possible. Concierges are often compared to magicians because they are able to procure the most basic or the most outrageous request—from airline reservations to hard-to-get concert tickets.

Although they have long been mainstays of many European hotels, it was not until the mid-1970s that concierges

became more prevalent in the United States. Today, concierges are found mostly in large urban hotels. Their offices range from massive counters equal to the front desk in size, or small niches arranged in one corner of the hotel lobby. Most often, they are found behind a large desk in the lobby, near the front desk. Armed with their computers, rolodexes, and telephones, concierges serve all hotel guests.

Many of the requests concierges receive are very basic, such as directions to city attractions, recommendations

---

To Be a Successful Hotel Concierge, You Should...

- be very organized and detail-oriented
- have the ability to juggle a number of duties and responsibilities at one time
- be courteous and cheerful with guests and coworkers
- be physically fit since you will spend much of your workday on your feet
- be discreet and respect the privacy of hotel patrons
- be able to handle sometimes hectic and stressful situations
- know the restaurants, tours, museums, and other attractions of the city or area you work in—or be willing to learn them

to tours or restaurants, or help dealing with airlines or car rental agencies. Concierges can take care of tasks such as obtaining out-of-town newspapers, dry cleaning, mailing packages, or reserving show tickets. Concierges also work with other departments in the hotel to prepare for large groups, VIP guests, or any guest that may have special needs. Sometimes welcome letters or baskets of fruit are sent to such guests by the concierge desk. However, their duties do not end there. Many hotels provide different levels of concierge service, depending on the type of guest. Penthouse guests can enjoy a private reception, with a separate concierge department to meet their needs, as well as serve them afternoon tea and hot hors d'oeuvres and drinks during the cocktail hour. Some concierge desks also host similar cocktail hours in the lobby.

Sometimes, a request can be more outrageous. Concierges have been known to plan large dinners or receptions on short notice, design entire travel itineraries complete with lodging and tours, rent airplanes or helicopters, secure the front row seats to a sold-out concert or sporting event, or even fill a room with flowers to set the mood for a marriage proposal. Concierges are trained to use their resources and contacts to serve the guest in whatever manner possible. They will, however, refuse to help the guest in any acts unlawful or unkind—no illegal drugs, prostitution, or practical jokes that may be deemed hurtful.

Many of the concierge's duties are performed on their own time, away from

the office. They must research any restaurant, tour, attraction, or store before comfortably recommending them to a hotel guest. In addition to visiting the facility, they must verify the price or entrance fee, insurance, and in the case of tours and shuttles, their operator's license. In order to keep current with new museum attractions, concerts, or restaurants, concierges read city magazines and newspapers, as well as hotel trade magazines.

Another important duty of the concierge is to maintain decorum. A good concierge is always well groomed and dressed neatly. They never gossip about the guests. If a guest has a strange request, the concierge should always be very discreet so as not to embarrass the guest or the hotel.

At larger hotels, concierges are supervised by *concierge managers*. Kristin Bodmer is the concierge manager at the Hyatt Regency Scottsdale Resort and Spa at Gainey Ranch in Arizona. "My main job duties include overseeing the daily function of the Concierge, Regency Club, and Golf Departments," she says. "I have 12 employees spread among the different outlets. On any given day I can do training, accounting, food service, tournament and event planning, scheduling, payroll, and other tasks, but my favorite part is getting to help our guests have the best experience while they are at our resort! I enjoy making a recommendation for a guest and having them come back and tell me how much they enjoyed it. My least favorite part is doing everything you can for a guest and

The Genesis of the Concierge

Concierges have been around since the Middle Ages. The term *concierge* is derived from the Latin term *conservus,* meaning "fellow slave." In the past, certain slaves were trained to assist their owners, often traveling ahead to make sure accommodations and food were in order. Holding the keys to the castle, these slaves became doorkeepers, and they were trusted to make sure everyone was safely locked in for the night. Throughout the years, as luxury hotels were built across Europe, it was necessary to install concierges to provide the same type of service for their guests.

having them still be displeased, but you can't please everybody all the time."

WHAT IS IT LIKE TO BE A HOTEL CONCIERGE?

Kathy Smith worked as a concierge at the boutique hotel Affinia Chicago until returning to school for her graduate education. "The decision to become a concierge, or guest service agent as I was titled, was a combination of luck and opportunity," she says. "After a long summer filled with interviews and job-hunting, Affinia Hotel finally 'knocked at my door.' I accepted the position because the job description fit my personality to the tee. It asked for someone who put the customer first and for someone who knew what it meant to have a real Chicago experience. I have lived in Chicago and the surrounding suburbs most of my life; I definitely knew what the heartbeat of this city was all about."

Kathy's primary job duty was to provide excellent customer service. "Being green and fresh out of school, Affinia provided the best learning experience of my life," she recalls. "The hotel was going through a renovation and my job was to be able to smooth the wrinkles, make the guests happy, and to do all that with a smile on my face. One of my secondary duties was to be a city guide. 'Where do I go?, What do I do?, How do I do it?, When do I do it?, and Why should I do it?' were questions I faced daily. This part of the job was my joy. It really didn't feel like a job. It was a pleasure hearing what a great time a guest had because of something I recommended. I recommended, scheduled, and found tickets for everything from restaurants to sporting events."

Kathy really enjoyed her time as a concierge. "Seeing the smiles on people's faces after coming back from a great time doing whatever you recommended was really rewarding," she says. "A secondary perk was that businesses around the city wanted us to recommend them so we constantly got free passes or got a first look at restaurants, night clubs, or shows. To be honest, for me at least, there weren't a lot of cons. The only con, perhaps, was that for some reason I couldn't make a guest happy. Usually those were the types who wouldn't be happy no matter what you did anyway."

DO I HAVE WHAT IT TAKES TO BE A HOTEL CONCIERGE?

"The number one rule for a concierge is to be selfless," says Kathy Smith. "The only concern you have is to make the guest happy. Another skill is to anticipate a guest's needs and wants. For example, if you know a guest is going to a show, also give them a list of restaurants within a walking distance, have the times they have openings for reservations, arrange transportation, and have a list of places open late night that are within walking distance. Lastly, you have to be a people person. Your job as a concierge is to make their stay as pleasurable as possible, and that may mean giving them a personable experience. Talk to the guest, know their name, and make them feel special."

Organization skills are also important. Concierges must maintain listings of a variety of contacts, from tour operators to food vendors, in computer databases. It is common for concierges to keep information cards on guests that frequent the hotel, noting their particular likes and dislikes, as well as special requests.

Knowing the city you work in is important. A concierge spends a considerable amount of time researching restaurants, tours, museums, and other city attractions. How can you recommend a particular boutique without knowing what exactly is sold? Concierges spend their off-hours trying the newest venues, often as guests of the establishment.

Does this sound like a fantasy career so far? Wait—there are downsides to being a concierge. Hard work is not always appreciated. You may have pulled out all the stops to plan a wonderful evening complete with dinner, dancing, and drinks at the most exclusive restaurant in town, as requested by a guest, only to have your suggestions lose to a simple meal served by room service.

Concierges spend much of the day on their feet, greeting guests, making phone calls, running errands, or doing whatever it takes to make things happen. Flexibility is imperative in this job since there is no such thing as "a typical day." A good concierge must be ready to deal with a single guest, or a group of 20, always in a cheerful and courteous manner. Situations can get hectic, especially when it's the middle of the tourist season, or the hotel is full of conventioneers. "You really have to want to put yourself out there and interact with people all day long," says Kristin Bodmer. "You are either made for it or you're not. You really can't be having a bad day or at least show that you're having a bad day when you're with the guests or it could severely impact their experience at your hotel."

HOW DO I BECOME A HOTEL CONCIERGE?
Education

High School

A well-rounded high school education is a good starting point for the career of concierge. Speech classes will help you learn how to effectively communicate with hotel guests and coworkers. The ability to write well is another skill important to future concierges. Direction cards, hotel communications, and welcome letters to VIP guests are just some examples of where writing counts. Fluency in another language, especially French, Spanish, or Japanese, can be extremely valuable when applying for a job.

Interested students should find part-time or seasonal work in order to gain working experience. Your part-time job need not be in a hotel (although that's the best place to make contacts); consider working in the customer service department of a department store. What about working as a junior assistant for a wedding consultant or party caterer? If you are always recruited to show out-of-town relatives the city sights, then at least get paid for your efforts by working for a tour company. "I recommend that high school students take up as many customer service-related fields as possible," says Kathy Smith. "Whether it is that weekend job at the Old Navy or serving the guests at a local Applebee's—those experiences will teach you how to deal with the public and how to make them happy."

Postsecondary Education

College degrees are not required of all concierges. In fact, many successful and established concierges have a variety of educational and employment backgrounds—from managers to artists to teachers. However, in today's competitive job market it pays to have an edge.

Some hotel associations, such as the Educational Institute of the American Hotel and Motel Association, offer certification classes for concierges. Only one program, however, is directly involved with the Les Clefs d'Or organization—the International Concierge Institute (ICI). The ICI, in partnership with the International School of Tourism, offers training and certification courses at their Ft. Lauderdale, Florida, or Montreal, Canada, campus. Graduates are given junior Les Clefs d'Or status.

Each 18-week program is divided into three modules. The first module, designed to introduce the student to the hospitality industry, has courses on tourism and hotel trends, guest service, and human relations, as well as concierge behavior and protocol. The second module consists of a nine-week internship under the tutelage of a Clefs d'Or member. The last module is an independent study language certification program. The workload of the ICI program is rigorous, and the requirements strict. Many interested students apply for the program, but only a few, those showing potential and aptitude to be a good concierge, are accepted. A degree from the ICI carries its weight—graduates from the ICI are often quickly hired by prestigious hotels around the world.

Certification or Licensing

Though not a requirement, certification is viewed by many as a measurement of professional achievement. The Educational Institute of the American Hotel and Lodging Association offers certification classes and organization activity level requirements. Only one program, however, is directly involved with the Les Clefs d'Or organization—the International Concierge Institute (ICI). ICI, in partnership with the International School of Tourism, offers training and certification courses on tourism and hotel trends, guest service, human relations, concierge behavior, and protocol, as well as a required language program.

High school juniors and seniors who are interested in working in the hospitality industry can take advantage of the Educational Institute of the American Hotel and Lodging Association's Lodging Management Program. The program combines classroom learning with work experience in the hospitality industry. Graduating seniors who pass examinations and work in the lodging industry for at least 160 hours receive the certified rooms division specialist designation. Visit http://www.lodging-management.org for more information.

Internships and Volunteerships

Internships and volunteer opportunities can give you a taste of the career without the pressures and obligations of a full-time job. The experience can also bridge your academic history with actual work experience, which will look very impressive on your resume.

WHO WILL HIRE ME?

Approximately 20,000 concierges are employed in the hotel and lodging industry. Besides hotels, concierges are found in some large apartment build-

for concierges. The National Concierge Association offers certification to members who meet educational, experience,

ings and condominiums. The concierge services provided at one Boston high-rise apartment include taking clothes to the cleaners, watering plants, and caring for pets when occupants are out of town. Large upscale department stores such as Nordstrom offer concierge service for its shoppers, from complimentary coat and package checking to restaurant and store information. Nordstrom concierges also offer tours of the store.

There is no typical path to this career. Some concierges have only a high school education. Being a graduate of a hotel or concierge program will, however, give you an edge in getting hired. Many schools with such programs, such as ICI, offer job placement services. Also, check hotel industry publications as they often post employment opportunities. Les Clefs d'Or takes great pride in the worldwide networking program it provides its members.

WHERE CAN I GO FROM HERE?

The position of concierge often is so interesting and challenging that many keep their jobs their entire working career. However, there are many opportunities for those who want to advance to other departments. Because a concierge's duties are very people-oriented, similar positions, such as a front desk manager, should be considered. If a concierge has a hotel-management or even a business degree, and working experience, as well as superior management skills, they could work for the position of general manager.

Related Jobs

- bell captains
- camp directors
- caterers
- cruise directors
- party planners
- personal assistants
- tour guides
- travel agents
- wedding consultants

The extremely ambitious can also consider starting their own concierge business. *Personal concierges* strive to be personal assistants to those too busy to organize their home or run errands. For a fixed price, personal concierges are responsible for a set of weekly duties; special requests, such as planning dinner parties or buying Christmas gifts, require extra payment. Since retainer fees for a personal concierge are steep—anywhere from $400 to $800 a month—people requiring such services are either extremely busy, very wealthy, or both.

WHAT ARE THE SALARY RANGES?

The U.S. Department of Labor reports that concierges earned salaries that ranged from less than $17,560 to $40,220 or more in 2008. Those employed in by hotels earned a mean annual salary of $27,180.

Did You Know?

Seventy percent of Les Clefs d'Or concierges in the United States are women, according to Les Clefs d'Or. In Europe the majority of members are men.

According to the ICI, expect to earn in the low- to mid-twenties as a new concierge. An experienced concierge, with Les Clefs d'Or status, working at a large urban hotel, can expect to earn around $50,000 a year. The usual benefits include paid vacations, sick and holiday time, health insurance, and some type of employee hotel discount, depending on the establishment.

A concierge, especially a good one, is often given tips or gifts by grateful hotel patrons. Ethically, concierges cannot and will not press for tips. The hotel guest decides whether to tip, and, if so, how much.

WHAT IS THE JOB OUTLOOK?

Employment for concierges employed in the hotel and lodging industry is expected to grow faster than the average for all careers through 2016, according to the U.S. Department of Labor. However, this is not a large occupation, and turnover is not high for these highly sought-after positions. "Unfortunately, in this economy," says Kathy Smith, "lei-sure and business travel is down, which means less business for the hospitality industry. However, we all know that when the economy goes down, it will also go up. We made it through post-9/11 when hospitality was at an all-time low, so my best advice for job seekers is to keep looking for jobs."

Employment opportunities are also available abroad, although European standards and training may be different from those found in the United States (check with the ICI). For those eager for challenges outside of the hotel industry, try cruise lines, rental properties, or consider setting up your own concierge business.

Busier lifestyles leave little time for mundane chores or last-minute details. Many travelers, once shy or intimidated by the concierge desk, realize that this is a free service available for their convenience. As people travel more frequently, they become more savvy. After spending hundreds of dollars for a double room, guests expect more than a bed and cable television. They are paying for a sense of luxury. Hotel general managers realize that a concierge department can provide the ultimate in guest services, and it is that type of service that makes a hotel a true luxury hotel. Even many smaller hotels, especially those that cater to business travelers, are now providing concierge service.

Some hotels have experimented with computerized kiosks that display tour and restaurant information. Though kiosks may cost less compared to maintaining a concierge department, they do not provide the one-to-one personalized service hotel guests desire. Also, kiosks cannot make a true recommendation since tour operators and restaurants pay a service fee to be advertised in the kiosk listing.

There are many stories about the crazy and fantastic requests concierges are asked to fulfill. Concierges who are able to meet the challenge are certainly imaginative and resourceful. However, it is not completing the near impossible that makes a concierge; rather, it is accommodating the simple requests for car rentals or directions to city museums that make a good concierge—one people trust and identify with true guest service.

Hotel Managers

SUMMARY

Definition
Hotel and motel managers are responsible for maintaining the daily operation, promotion, and policy of their lodging establishment. They oversee all staff activities, services, budgeting, buying, sales, and security.

Alternative Job Titles
Lodging managers

Salary Range
$28,160 to $45,800 to $84,270+

Educational Requirements
Associate's degree; a bachelor's degree in hotel management is increasingly recommended

Certification or Licensing
Recommended

Employment Outlook
About as fast as the average

High School Subjects
Business
Mathematics
Speech

Personal Interests
Business management
Helping people: personal service
Selling/making a deal

It takes not only good organization and leadership skills to be a successful hotel general manager, but also the ability to nurture, support, and develop your staff. Tom Pagels, general manager at Hyatt Regency McCormick Place, had such an experience with an executive steward named Robert, whom he worked with in his first assignment as general manager.

"Robert came from a family that was broken by drugs and criminal acts committed by other family members," Tom recalls. "He didn't have any good reason to expect much of himself. When I arrived at the hotel, he was not faring very well. Most of his challenge was for lack of training. No one had ever taken time to simply show Robert the ins and outs of his profession. This led him down roads of confrontation with other managers and left an impression of incompetence regarding his performance.

"Over a period of several months, Robert and I struck up a relationship that still remains today. Ultimately Robert was honored as department head of the quarter for his contributions in the food and beverage division of the hotel. Later, following my move from the hotel, Robert was promoted to banquet manager.

"This was likely the most rewarding experience of my career: to see someone not only pass my expectation, but clearly surpass his own."

WHAT DOES A HOTEL MANAGER DO?

It takes a variety of specially trained managers to keep a hotel running smoothly. The following paragraphs detail some managerial specialties in the hospitality industry.

A *hotel general manager* is like the ringmaster of a circus. He or she has to manage several tasks simultaneously, giving each department individual, but equal, attention, all the while making sure the guests are happy and satisfied. Does this sound like an easy job? Think again.

General managers are responsible for the overall supervision of the hotel, the different departments, and their staff. They follow operating guidelines set by the hotel's owners, or if part of a chain, by the hotel's main headquarters and executive board. The general manager allocates funds to all departments of the hotel, approves expenditures, sets room rates, and establishes standards for food and beverage service, hotel décor, and all guest services. General managers tour the hotel property every day, usually with the head of the housekeeping department, to make certain the hotel is kept clean and orderly. General managers are responsible for keeping the hotel's accounting books in order, advertising, room sales, inventory, and ordering of supplies, as well as interviewing and training of new employees. However, in larger hotels, the general manager is usually supported by one or more assistants.

Some hotels still employ *resident managers*. Such managers live at the hotel and are on call virtually 24 hours a day, in cases of emergencies. Resident managers work a regular eight-hour shift, attending to the needs of the hotel. In many modern hotels, the general manager has taken the place of resident managers.

Front office managers supervise the activities and staff of the front desk. They are responsible for directing reservations and sleeping room assignments. Front office managers make sure that all guests are treated courteously, and that check-in and check-out periods are managed smoothly. Any guest complaints or problems are usually directed to the front desk first—managers are responsible for rectifying all criticisms before they reach the general manager.

The *personnel manager* heads the human resources or personnel department. He or she is responsible for hiring and firing employees and works with other personnel employees, such as employee-relations managers, to protect employee rights and address grievances.

Executive housekeepers are managers who supervise the work of the room attendants, housekeepers, janitors, gardeners, and the laundry staff. Depending on the size and structure of the hotel, they may also be in charge of ordering cleaning supplies, linens, towels, and toiletries. Some executive housekeepers may be responsible for dealing with suppliers and vendors.

Restaurant managers oversee the daily operation of hotel restaurants. They manage employees such as waiters and waitresses, busboys and -girls, hosts and hostesses, bartenders, and cooks and bakers. They are responsible for customer complaints and satisfaction.

Food and beverage managers are responsible for all food service operations in the hotel, including restaurants, cocktail lounges, banquets, and room service. They supervise food and service quality and preparation, order supplies from different vendors, and estimate food costs.

A *security manager*, sometimes known as a *director of hotel security*, is entrusted with the protection of the guests, workers, and grounds and property of the hotel.

Hotels can profit by marketing their facilities for conventions, meetings, and special events. Such hotels may have a *convention services manager* in charge of coordinating such activities. He or she takes care of all necessary details, such as blocking sleeping rooms and arranging meeting rooms or ballrooms. The convention services manager is responsible for resolving any problems that arise.

Hotel chains employ specialized managers to ensure that its hotels are being operated appropriately and in a financially sound manner. *Regional operations managers* travel throughout a specific geographic region to see that hotel chain members are operated and maintained according to the guidelines and standards set by the company. *Branch operations managers* reorganize hotels that are doing poorly financially, or set up a new hotel operation.

WHAT IS IT LIKE TO BE A HOTEL MANAGER?

Tom Pagels is the general manager at Hyatt Regency McCormick Place in Chicago. He has worked in the industry for approximately 33 years, and with Hyatt Hotels & Resorts for 29 years. "I clearly remember the moment I embarked on the idea of hospitality," he recalls. "I was sitting in a second-year contract law class bored totally out of my mind. It became painfully obvious that I needed to set a course. With only sports and partying on my mind, I knew I would never become a professional athlete, so party planning was the next great passion. The more I worked in the industry, starting as a bartender and then a cook, it became clear that hotels were the direction I should take."

Tom says he likes the fact that his primary duties as general manager change as the demands on his hotel property change. "My duties include service and customer satisfaction, sales of rooms, food and beverage, and, of course, profit maximization," he says. "Within these primary objectives exist an endless variety of challenges that keep the job fresh from day to day or year to year. It is, in fact, why I remain motivated and challenged to be the best in this industry at what I do.

"What I like most is that this business is all about people—whether it's motivating, servicing, celebrating, or counseling," Tom continues. "There is something gratifying about seeing people aspire to levels they didn't even believe they could achieve. Or when a guest is so totally taken by the manner in which they are treated. To break this business down, it is people providing service to others. I cannot imagine a more challenging scenario."

Ethan Shelton is the assistant executive housekeeper at Hyatt Regency Washington on Capitol Hill in Washington, D.C. He has been employed at this Hyatt property for almost seven months, but he began his career with Hyatt in 2005 when he was hired for a bell attendant position at the Hyatt Regency in Columbus, Ohio. He also worked as a front office agent and as a housekeeping supervisor. "I couldn't tell you how many times I changed my mind when deciding what I wanted to do with my life (career-wise)," Ethan says. "I always thought the idea of owning a hotel would be the coolest thing in the world; this was inspired by the Hyatt Regency Indianapolis, with its gigantic atrium lobby and a mall located on the ground level. I went there for a family reunion when I was seven. When taking an exploration class at Ohio State University I heard about its hospitality management program and knew that's where I belonged. The hospitality industry is extremely broad, and it is one of the largest industries in the world, if not the largest. The summer after my fresh-

man year in college I needed money and wanted to get my feet wet, so I applied to the best hotel name I knew of—Hyatt.

"I am responsible for assisting in the overall success of the housekeeping department," he continues. "The success can be measured by three main components: financial management, internal guest scores, and external guest scores. Some secondary job duties include managing email (effectively), coaching and assisting associates, decision making, presence in the public areas and guest rooms, and purchasing supplies and equipment. Do I have to get 'down and dirty' and clean sometimes? Absolutely, anybody going into my field should be ready to do so!"

Ethan says that one of the best parts of his job is that every day is different. "I deal with a lot of different people and am placed in different situations every day, so it never gets boring," he says. "I also get satisfaction when I find that I have made somebody happy, which is what our company is all about, both internally and externally. On the downside, we are a business, and businesses must be competitive to survive and prosper. This can be challenging when it comes to working long hours and managing stress. Secondly, I do feel that with as hard as a lot of people work in my field, there are many other fields that pay much better."

Ethan says that too many interesting things have happened to him during his career in hospitality to cite just one event. "Our world is amazingly large and there are infinite things to explore, learn about, and experience," he says. "With my career I have been able to travel to neat places, move around, learn about the coolest things, and experience some awe-inspiring events. I will never forget when I was in Chicago for Hyatt's !mpact Training, and a group of us got to go downtown for President Obama's acceptance speech in Grant Park. I also got to be in the heart of President Obama's inauguration (even though I was working). Not many people were able to be at both events without paying hundreds or thousands of dollars; I did not pay anything except a few train fares."

Walter Stanley is the executive assistant manager of rooms at the Hyatt Regency Bethesda in Bethesda, Maryland. He has worked in the hospitality industry for more than 20 years, with 14 years dedicated to Hyatt. "The Hyatt Corporation gave me the opportunity immediately following college," Walter says, "and has been the major contributor to my growth both personally and professionally. The thought to enter the hospitality field began when I was very young. As a child I had the opportunity to travel cross-country via automobile with my family, and during these trips we stayed in several establishments—from motels to high-end hotels. This introduction to hotels made me want to become a part of something that developed my vision of service and what it felt like to make guests happy.

"My primary responsibilities," he continues, "are to manage the Rooms Division (Reservations, Front Office, Communications, Guest Services, Housekeeping, Security) and focus on how we can improve the engagement between associates and guests. Secondarily, I am responsible for the training and support of all associates within my division. I directly report to the general manager daily and assist with executing the company's philosophy. The most rewarding thing that comes from this industry is when you look into a guest's eyes who never has been to your city and you make their vacation. All of the promotions and rewards are great, but seeing exceptional guest satisfaction at its finest, there is nothing greater."

Jackie Grabow is the director of activities at the Hyatt Regency Scottsdale Resort and Spa at Gainey Ranch in Arizona. She has worked in the hospitality industry for seven years, with four of these employed for Hyatt Corporation. "I decided that I wanted to work in hospitality," she says, "because I love working with travelers of diverse populations, I love the excitement of having new challenges each day, and I love being involved in providing a quality experience that travelers will cherish in their memory for years to come. Some of my duties as a manager include hiring and training associates, managing associates' performance, planning activities for Camp Hyatt and families at the resort, representing my department to groups at preplanning and pre-convention meetings, maintaining a pristine facility, arranging department meetings, setting up relationships with vendors and handling purchasing for our department, maintaining a safe work environment by observing potential hazards, and creating a fun and friendly environment for my staff and my guests. I love that I have many different experiences at work every day, I get to perform a wide variety of duties, keeping my job interesting."

Jackie says that it is sometimes difficult to work on every holiday. "Working holidays and weekends is something that comes with the territory," she says, "and you must be prepared to do so. At the same time, I get to create an unforgettable holiday experience for our guests, and that is quite rewarding.

"To be in this career," she continues, "you must be a jack of all trades, you must be flexible, and you have to be hospitable. I have had families that come back to my property every Easter or every 4th of July and rave about the activities and how it 'brings the family together.' I have to say it is extremely rewarding to have kids remember your name from their visit last year; when they see you they run to give you a hug or a high-five. That really makes me feel like I've created something special!"

DO I HAVE WHAT IT TAKES TO BE A HOTEL MANAGER?

Hotel managers are strong leaders with a flair for organization, communication, and, most importantly, working well with other people. If this description fits your personality, great; if you lack two or more of these traits, then maybe this is the wrong career path for you.

Good managers should initiate, implement, and praise the work of their staffs. In order to keep the hotel running smoothly, managers need to establish policies and procedures and make certain their directions are carried out. They must be able to solve problems and concentrate on details, whether large or small. It is a stressful job and managers need to keep a cool demeanor when dealing with difficult situations that come their way. Managers must be born diplomats, especially when handling guest complaints. They need to validate all criticisms, no matter how trivial, and find the fastest and most satisfying solution to the problem.

"Today, without a doubt, the most important quality for a hotel manager is inclusion," says Tom Pagels. "If a person is going to choose this line of work, they are required to deal with all types of people. This can be very rewarding and inspiring. For a person who is open minded and willing to listen to various opinions, you will likely be pleasantly surprised by the free exchange of ideas and results that follow. Today, everyone is privileged to have information flowing at incredible rates in all avenues—cutting edge and traditional. As a result, everyone has an opinion and is ready to share. If this is embraced, it can be very powerful in moving an organization toward success."

To Be a Successful Hotel Manager, You Should...

- have strong leadership, organization, and communication skills
- be able to work with different types of people and situations
- be a good problem-solver and be attentive to details
- be a born diplomat, especially when handling guest complaints
- be willing to work long hours, including late nights, weekends, and holidays

"In order to be successful in my career you have to be personable," says Ethan Shelton. "You also need to genuinely care for people and have a sense of pride for the product you are selling. We are not only selling a physical product (the guest room, food, meeting space, etc.), but we are selling a service, and *you* have to make it genuine and exceptional. This career requires a good combination of management and leadership skills. Not only should you be organized and educated about how to work with people (in today's diverse atmosphere), but you must be able to look beyond the current moment, be proactive, and be inspirational."

Do you hope to have a nine-to-five workday? You won't have one with this career. Managers usually work an average of 55 hours a week. Weekends and

holidays are no exceptions. Even if off duty, managers can be called back to work in cases of emergency—night or day.

HOW DO I BECOME A HOTEL MANAGER?
Education

High School

It's a good idea to begin preparing for a career in hotel management while in high school. Concentrate on a business-oriented curriculum, with classes in finance, accounting, and mathematics. Since computers are widely used in the hotel setting for reservations, accounting, and management of supplies, computer literacy is important. Brush up on your speaking skills while in high school—you'll need them when giving direction and supervision to a large and diverse staff. A second language—especially Spanish, French, or Japanese—will be very helpful to you in the future.

Instead of working at the mall or local fast-food outlet, play it smart and consider obtaining part-time or seasonal work at a local hotel. Some hotels employ high school students to work the switchboard and front desk, and as members of banquet, restaurant, and housekeeping staff. In addition to making money, these teenagers are getting valuable experience and making useful contacts for future employment. Check with your high school career center to see if there are employment opportunities in your area. Don't forget that career centers, your local library, and the Internet can

Hotel Manager Spotlight: Joanna Bentz

Joanna Bentz is the assistant food and beverage manager at the Hyatt Regency Washington on Capitol Hill in Washington, D.C. She has worked in the field for 10 years. Joanna discussed her career with the editors of *What Can I Do Now? Travel and Tourism.*

Q. Why did you decide to enter this career?

A. I entered this career by force—not choice—is what my parents like to remind me. I began in grade school at church at the front doors greeting guests as they walked in for service. I was put to work at church because I had a tendency of finding trouble when I was younger. To help focus, I was kept busy at church and soon found that I really did enjoy helping in the community service projects and services I participated in at church. When I turned 15 and was allowed to work, I obtained a job where I worked the host stand and for the following six years moved to different positions throughout the company. Throughout college I had different jobs but all related back to hospitality being a server, bartender, or greeter. After college I started my job with Hyatt Hotels & Resorts. I stayed in this field because it's what I do. It's second nature to me to give people what they want the best way I know how, or to make their experience, time, or moment the best it can be.

Q. What are your main and secondary job duties?

A. My main job duty is to assist in overseeing the outlets on property (outlets being the lounge, restaurant, and room service), and supporting the staff and ensuring they have what they need to take care of the guests in a timely manner. Secondary job duties include more administrative duties, such as schedules, payroll, allocating tips, completing reports, and so forth.

Q. What do you like most and least about working in this field?

A. What I like the most is that no day is ever like another. You may have an idea of what your day is going to be like, but you never know when and with whom situations will and can come up. I like the variety of the field as well—from the people, to the seasons, to the groups that come to the property, to the employees who work here. It's never a dull day in food and beverage! I wouldn't say there is anything I like least because it's a give and take for all things in hospitality. You may not have a holiday off when the rest of your family does, but you do have time off during the week when most people are at work and you can get your errands done without a ton of people in your way.

Q. What are the most important personal and professional qualities for people in your career?

A. Most important, both personally and professionally, is the ability to keep a cool head in all situations. Many situations arise, and not having a level head to figure out the next step can be detrimental to the operation and

(continued on next page)

(continued from previous page)

the guests' experience. The ability to multitask is also a necessity because so many things are going on at once. Multiple things need to happen at once, and successfully. An important personal quality would also have to be one's passion for this field. You can only be so successful if you are showing up for a job for the paycheck, but if you are showing up with a passion for the job, then it will make your work that much more rewarding for you.

Q. **What is one of the most rewarding things that has happened to you while working in this field?**
A. The most rewarding thing that has happened is how much this position has humbled me. Being able to understand that there is a lot of room to grow and learn, regardless of being in a classroom or not, is very valuable. I have found it most interesting that so many people stay in this field for so long because they really are passionate about their guests and their jobs.

all be helpful when researching college programs or specific businesses.

Postsecondary Training

Many companies require management trainees to have a minimum of a bachelor's degree in hotel and restaurant management. There are more than 800 colleges and universities that offer undergraduate and graduate programs in hotel administration. Visit the American Hotel and Lodging Association's Web site (http://www.ahla.com/products_list_schools.asp) for a list of colleges and universities that offer degrees in hospitality and hotel management.

A typical hotel management program will concentrate on hotel administration, food service management, accounting, economics, marketing, housekeeping, computers, and hotel maintenance engineering. To complement class instruction, most programs require students to work on site at a hotel. Some universities, such as the famed Cornell School of Hotel Administration (http://www.hotelschool.cornell.edu), have a training hotel on campus. Many hotels will also consider candidates with degrees in business management, public relations, or other related fields, if they are highly qualified and talented.

Certification or Licensing

Certification for this job is not a requirement, though it is recognized by many as a measurement of industry knowledge and job experience. The American Hotel and Lodging Association (AHLA) Educational Institute offers a variety of certifications for hotel and motel managers, including certified hotel administrator, certified lodging manager, certified lodging security director, certified food and beverage executive, certified hospitality housekeeping executive, certified human resources executive, certified engineering operations executive, certified hospitality revenue manager, and certified lodging security supervisor. Contact the institute for more information on these certifications.

Additionally, the International Executive Housekeepers Association offers two certification designations: certified executive housekeeper and registered executive housekeeper. Qualified individuals earn these designations by fulfilling educational requirements through a certificate program, a self-study program, or a college degree program. Other hospitality trade associations also offer certification programs for hotel managers.

High school juniors and seniors who are interested in working in the hospitality industry can take advantage of the Educational Institute of the AHLA's Lodging Management Program. The program combines classroom learning with work experience in the hospitality industry. Graduating seniors who pass examinations and work in the lodging industry for at least 160 hours receive the certified rooms division specialist designation. Visit http://www.lodging management.org for more information.

Internships and Volunteerships

Participating in internships and volunteer opportunities will provide you with a great introduction to this career. These activities also look great on a resume and may give you an edge over other job applicants. The best sources of information about such opportunities are professional organizations, such as the Educational Institute of the American Hotel and Lodging Association.

Your high school or college career center will have helpful information on available internships or volunteer opportunities, as well as career books, magazines, and counselors to guide you. Don't forget to ask your teachers if they know of job opportunities before the center does. Post your resume on an industry job Web site or research the countless internships and scholarships that are posted by schools, hotels, and motels across the world.

WHO WILL HIRE ME?

"I started in the hospitality industry by working in a small resort in northern Michigan called Shanty Creek," recalls Tom Pagels. "I was a bartender and learned all the particulars of running the bar from an opening, closing, entertainment, and staffing perspective. This was a lot of fun and it kindled my passion for the industry. After a year, I became opening breakfast cook, and the responsibilities of that position included making the soups and sauces of the day. This was really good practical training, just what I needed. It was after a year or so when I was getting anxious to be in charge or see the world. Both options were exciting, but I decided to try to see the world. This ended with a stint in the Bahamas working for the Windjammer Barefoot Cruises. That sealed the deal

Hotel Manager Spotlight: Donna Lowell

Donna Lowell is the director of human resources at the Hyatt Regency Scottsdale Resort and Spa at Gainey Ranch in Arizona. She discussed her career with the editors of *What Can I Do Now? Travel and Tourism.*

Q. Why did you decide to enter this career?

A. I was retired as a human resources director for a telephone company and wanted to relocate back to Boston. There was a human resources director position open at one of my favorite hotels. I thought it would be a great place to work and use my experience and skills in human resources.

Q. What are your main and secondary job duties?

A. My primary job duties include running a department of three managers (employment, training and payroll/workers compensation, benefits and recognition). My secondary job duties include facilitating and training our Green Team on our Hyatt Earth Project, assisting managers in developing their future career path, and serving as a certified tourist ambassador for the Scottsdale Convention and Visitors Bureau.

Q. What do you like most and least about working in this field?

A. I definitely enjoy the spontaneity of the industry; anything can happen at anytime. The perks of being able to stay at another Hyatt Hotel for free or at discounted rates are tremendous, especially if you like to travel. I love meeting new people since I'm very outgoing and being involved in the overall hotel operation. What I like the least is the number of hours you need to sometimes work. It can be a demanding industry because it's 24/7.

Q. What are the most important personal and professional qualities for people in your career?

for me, and I finished school to join Hyatt in 1981."

"My training began by taking any position available to get my foot in the door," says Walter Stanley. "I knew with my passion for the industry it only took an opportunity, and I could excel from there. After working on the food and beverage side of the industry, and attending the University of Maryland, Eastern Shore to hone my skills, I truly knew this was the industry that would help me build a career and not just a job. I continued to work diligently to learn all I could and stayed focused on what got me interested in the industry—service."

There are approximately 71,000 hotel and motel managers working in the United States. Unless you are lucky to be part of a family that is already in the business, you'll have to do some future planning and maybe pound the pavement a little. If you are looking for a job to lay the groundwork for college, then

A. The most important qualities personally would be a great personality and attitude (then you can train the rest). You need to be a people person and want to provide great service to guests. We look for someone who would like to make it a career (although we do have many associates working part time and attending college). Professional qualities vary from position to position. We may require licensing for the spa positions as well as an engineering degree or certification to work in facilities. You would need to possess some management experience if you wanted to begin here as a manager. Hyatt also has a tremendous Corporate Management Training Program for college graduates. For a human resources director position, candidates are expected to have already had experience with employment, benefits, employee relations, and compliance and have a direct knowledge of federal laws and some state laws. Certification in the human resources field is a plus, but not required. It's important that you have a working knowledge of the hotel operations and excellent critical thinking skills. With many facets of human resources being outsourced to other companies (such as benefits, payroll, and leaves of absence) it is important for a human resources director to be a strategist and have the ability to assist other operational directors with organizational development to streamline overall hotel operations.

Q. What is one of the most interesting or rewarding things that happened to you while working in this field?

A. Most interesting would be some of the celebrities I have met along the way and what they like and dislike (food, amenities, room requirements) when they stay at our hotel. Every day for me is rewarding, especially when I have an associate come to me and thank me for assisting them (with their benefits, career development plan, etc.). Also, it is rewarding when we get the results of the employee survey and it reflects that they are happy with the company, their position, and how they are treated.

start with your school's career center or your guidance counselor. Also, consider looking in your community paper under "Hotel," "Motel," "Restaurant," or "Business." It may be a good idea to call local hotels, or even restaurants, to see if they are hiring for seasonal work. Hospitality job listings are also available online.

College seniors can make job inquiries with their school's career services office. Usually such centers will try to match a student's skills and education with the right position. Many schools also hold job fairs where prospective employers can set up immediate interviews with qualified candidates. Make sure you dress professionally and have copies of your resume ready to circulate. Not all companies send recruiters to campus job fairs, in which case you can send your resume directly to their human resources department. Make the extra effort of finding out who is

> ## Advancement Possibilities
>
> *Branch operations managers* reorganize hotels that are doing poorly financially, or set up a new hotel operation.
>
> *Owners of an inn or bed and breakfast* establishment are called *proprietors*. They are responsible for every aspect of the business—from setup, advertising, and ordering supplies and food to guest relations.
>
> *Regional operations managers* travel throughout a specific geographic region to see that hotel chain members are operated and maintained according to the guidelines and standards set by the company.

in charge of hiring, so your cover letter won't read "Dear Sir or Madam."

Some major employers in the industry are Wyndham Worldwide (Days Inn, Super 8, Ramada, Howard Johnson, Travelodge), Inter-Continental Hotels Group (Holiday Inn), Hilton Hotels Corporation (Doubletree, Embassy Suites, Hilton), Global Hyatt Corporation (Hyatt, Hyatt Regency, Hyatt Resorts, Grand Hyatt, Park Hyatt, Hyatt Place, Hyatt Summerfield Suites, and Andaz), and Choice Hotels International Inc. (Comfort Inn, Econo Lodge, Sleep Inn). These companies have properties located nationwide and abroad. Marriott International Inc., another international player, offers a fast-track management program for qualified employees and has been known to encourage career advancement for minorities and women.

WHERE CAN I GO FROM HERE?

The position of manager is among the top rungs of the hotel career ladder. It's unlikely this would be your first industry job. In the past, employees could advance from the front desk, housekeeping, or food and beverage departments, or even the sales staff. Good experience and hard work was sufficient to move ahead. However, the current reliance on computers and technology coupled with a competitive work force make experience alone insufficient for job advancement. Most candidates have some postsecondary education; many have at least a bachelor's degree in hotel and restaurant management. Graduates entering the hotel industry usually pay their dues by working as assistant hotel managers, assistant departmental managers, or shift managers. Many hotels, such as the Marriott or Hilton hotels, have specific management training programs for their management-bound employees. Employees are encouraged to work different desks so they will be knowledgeable about each department of the hotel.

The average tenure of a hotel general manager is about six and a half years; those who have worked as a general manager for 10 years or more usually view their job as a lifetime commitment. Managers who leave the profession usually advance to the regional or even national area of hotel management, such as property management or the administrative or financial positions of the hotel chain. Some may opt to open their own hotel franchise, or even operate a small inn or bed and breakfast establishment. The management skills learned as a general manager are valuable and can be successfully utilized in any avenue of business.

WHAT ARE THE SALARY RANGES?

Salaries vary according to the worker's level of expertise, the lodging establishment, the duties involved, and of course, the location of the hotel. According to the U.S. Department of Labor, lodging managers reported a median yearly income of $45,800 in 2008. The lowest paid 10 percent earned less than $28,160 annually, and the highest paid 10 percent made more than $84,270 per year.

All managers receive paid holidays and vacations, sick leave, and other benefits, such as medical and life insurance, pension or profit-sharing plans, and educational assistance.

WHAT IS THE JOB OUTLOOK?

Employment for lodging managers is predicted to grow about as fast as the average for all occupations through 2016, according to the U.S. Department

Related Jobs

- apartment house managers
- aquatic facility managers
- building superintendents
- camp directors
- casino managers
- golf club managers
- health care managers
- health club managers
- hospital administrators
- recreation center managers
- restaurant managers
- travel agency managers

of Labor. However, the number of jobs for hotel managers is not expected to grow as rapidly as in the past. Travelers are now looking for bargains and a no-frills type of lodging. To meet the demand, many hotel chains are offering a larger choice of economy properties, which as a rule do not have many of the amenities found in luxury hotels, such as in-room food and beverage service and in-house restaurants. Because there are fewer departments, fewer managers are needed to manage these hotels.

Newly built hotels and luxury or resort properties will continue to need well-trained and experienced managerial employees. Many job openings will also result from current managers moving to other positions or occupations, retiring, or leaving the work force for other reasons. Employment and advancement opportunities will be best for those with college degrees in hotel or restaurant management, or a similar business degree. Managers with excellent work experience will be in demand, as well as those who have successfully completed certification requirements. As hotels and motels become more computer-dependent, familiarity with different computer software programs will be key.

Pilots

SUMMARY

Definition
Pilots who work in travel and tourism fly aircraft that carry passengers to and from various destinations. Other pilots may work for governmental agencies, medical facilities, or be involved in transporting cargo or performing agricultural work.

Alternative Job Titles
Aircraft pilots
Airline pilots

Charter pilots
Commercial pilots
Private pilots

Salary Range
$55,330 to $111,680 to $200,000+

Educational Requirements
Bachelor's degree, flight training

Certification or Licensing
Required

Employment Outlook
About as fast as the average

High School Subjects
Mathematics
Physics
Shop (trade/vo-tech education)

Personal Interests
Airplanes
Figuring out how things work
Fixing things
Travel

It's a beautiful spring morning. The air is clean and crisp and there isn't a cloud in the sky. Andrew Targosz is at the controls of a plane that is gently lifting into the air. He thinks back to when he was a child and dreamed of being a pilot. Now, he does it every day—and gets paid for it.

More importantly for Andrew, he's doing something he really enjoys. His job of working for a major commercial airline is interesting and never boring; he never knows what to expect from week to week, or even day to day.

WHAT DOES A PILOT DO?

Pilots who work in travel and tourism fly aircraft that transport passengers. While there are several other kinds of pilots— such as agricultural pilots, military pilots, helicopter pilots who fly for law enforcement agencies and hospitals, and pilots who fly cargo planes—this chapter will focus on pilots who fly commercial and charter aircraft.

Commercial pilots, who fly for major airlines such as United and American, are the best-known and largest group of professional pilots. There are two main designations of commercial airline pilots: *captain* and *copilot*. The captain is usually the pilot with the most seniority. He or she is in charge of the plane, with the copilot being second in command.

Aside from actually flying the aircraft, pilots have a variety of safety-related responsibilities. Before each flight, they must determine weather and flight conditions, ensure that sufficient fuel is on board to complete the flight safely, and verify the maintenance status of the aircraft. They also perform a system operation check to make sure that all instrumentation, controls, and electronic and mechanical systems are functioning properly. Before the plane takes off, the captain briefs other crew members, including flight attendants, about the flight specifics.

After all the preflight duties have been performed, the passengers have boarded, and the flight attendants have secured the cabin for takeoff, the captain taxis the aircraft to the designated runway. When the control tower radios clearance for takeoff, he or she taxis onto the runway and begins the takeoff. As the plane accelerates for takeoff, the captain focuses on the runway, while the copilot monitors the instrument panel. To determine the speed needed to become airborne, the pilot must factor in the altitude of the airport, outside temperature, weight of the plane, and wind direction.

Except for takeoff and landing, most of the time a large commercial jet is in the air, it is actually being flown on autopilot, a device that controls the plane's course and altitude, making adjustments to keep it on course. Planes today may even land on autopilot. This does not mean, however, that pilots can sit back and relax. They must constantly monitor the aircraft's systems and the weather. They also remain in constant communication with air traffic controllers. As the plane travels, the pilots communicate with a series of radio navigation stations along the route.

When the cloud cover is low and visibility is poor, pilots must depend completely upon instrumentation. Altimeter readings tell them how high they are flying and whether they can fly safely over mountains and other obstacles. GPS technology provides pilots information that, coupled with special maps, tells them their exact position.

As the plane nears its destination airport, the pilot radios the air traffic controller for clearance to approach. To land, the plane must be maneuvered and properly configured to make a landing on the runway. Once he or she has received clearance and positioned the plane for landing, the pilot extends the landing gear into the down position and sets engine power for the approach. After the plane touches down, the pilot taxis the plane to the ramp or gate area where the passengers deplane.

Pilots keep detailed logs of their flight hours, both for payroll purposes and to comply with Federal Aviation Administration (FAA) regulations. They also follow "afterlanding and shutdown" procedures, and inform ground maintenance crews of any problems noted during the flight.

Flying for a large commercial airline carries much responsibility. The aircraft itself costs millions of dollars, and the safety and welfare of dozens to hundreds of passengers are on the line each time a plane makes a flight. All commercial pilots must undergo continuous testing and evaluation to make sure their skills are in top shape. Each major airline has its own testing requirements, but most of them involve annual or semiannual testing of each pilot's ability. Flying and navigating are considered primary flying responsibilities. Secondary flying responsibilities include filing flight plans and listing flight reports for the FAA.

FAA regulations limit airline pilots to no more than 100 flying hours per month (or 1,000 hours a year). Most airline pilots fly approximately 65–75 hours per month and spend another 65–75 hours a month on other duties.

Charter pilots have essentially the same job that commercial pilots do, but on a smaller scale. Because they work for a much smaller organization with far fewer employees, charter pilots usually have more secondary flying responsibilities than commercial pilots. They may also be involved in such tasks as loading and unloading baggage, supervising refueling, keeping records, scheduling flights, arranging for major maintenance, and performing minor maintenance and repair work.

WHAT IS IT LIKE TO BE A PILOT?

Andrew Targosz has worked as an airline pilot for a major airline carrier for the past 24 years. "I entered this field because

---

Lingo to Learn

cockpit The area in the front of the plane where pilots sit. Flight controls and instruments are located here.

flight hours Term used to describe the amount of flight time a pilot or potential pilot has accumulated. To obtain certain licenses, pilots need to fly a certain number of hours.

flight school Place where prospective pilots gain flight instruction, both in the air and on the ground. Flight schools can own or lease their planes.

instrument panel The area directly in front of the cockpit where instruments such as the altimeter, air speed indicator, and fuel gauge are located.

instrument rating Term used to describe advanced certificates pilots may earn for completing additional training that teaches them to pilot a plane by using instruments only, as opposed to visuals (ground, landmarks, etc.).

simulator Device used to test pilot's flight knowledge. Creates artificial flight circumstances.

I've always had a love of flying," he says. "Currently, I fly Boeing 757s and 767s for continental and international flights. When I started, I had no real aspirations to become an airline pilot. I just liked airplanes, and I liked flying."

Andrew says that one of the best things about being a pilot is the feeling of freedom flying in the clouds. "Additionally, with international travel, I have the opportunity to travel abroad. I also like the fact that I'm not stuck in an office."

Although he loves being a pilot, Andrew says that there are a few drawbacks to the field. "First, the amount of different time zones you fly through is really hard on your body," he says. "Second, by law, you are limited to 100 hours of flying time per month, or 1,000 hours a year. However, the way a lot of schedules are written, you won't get the maximum 100 hours flying, but will still be away from home between 300 to 400 hours a month. I might have one flight from Chicago to New York, and then I sit all day before my next flight. I have only a small number of one-day trips. Third, you have more time off, but also have more time away from home. For example, you may have 10 or 12 days off a month, which seems like more than usual, but really there may be stretches of working days away from home. It can be hard on your family."

One of Andrew's most rewarding experiences came on a flight out of Los Angeles. "We ran into some trouble," he recalls. "We lost oil quantity and had to shut an engine down. Luckily we have simulator tests every nine months to give us practice with single-engine emergency situations. With 180 passengers on board, it was very rewarding to land and get them home safely."

Andy Simonds is a first officer (copilot) for American Airlines. He received his private pilot certificate in 1980 and, after earning additional certificates and ratings, started flying professionally in 1987. "When I decided to become a pilot," Andy says, "it was driven by one main factor: I wanted to enjoy what I did for a living. I wanted to be excited about where I was going when the alarm clock woke me up in the morning. I was interested in developing and honing a skill that was challenging and important. It also had to be fun. While this career has many challenges, flying airplanes around the world fits these criteria. That's not to say that it's all been easy or without disappointment, but the career has exceeded my expectations."

Andy has been flying for American Airlines since 1998 and is currently qualified to fly the Boeing 757 and Boeing 767 aircraft. "The 757," he explains, "is a narrowbody aircraft (with one aisle in the middle) that seats 188 passengers. The 767 is a widebody aircraft with two aisles. It holds 225 passengers. I have also flown the DC-9, Boeing 727, and various turboprop and single engine aircraft. Typically, airline pilots are qualified on only one type of aircraft at a time. The 757 and 767 cockpits and systems are similar enough that the FAA considers these two models as the 'same' type.

"Due to the fact that I am based in Boston as an international first officer on the 757/767 with American Airlines," he continues, "I may fly to London, Paris, the Caribbean, Central America, and South America. We typically fly 15 days a month, and our schedule changes (along with many of the trips) on a monthly basis."

Andy says that being a pilot isn't for everyone. "Obviously, a pilot is away from home a lot," he says. "Yes, we get to go to some nice places but before I could regularly fly to London or Paris, I had to spend a lot of time flying to Tulsa, Tuscaloosa, and the Tri-Cities [Kingsport, Tennessee; Johnson City, Tennessee; and Bristol, Virginia]. That was fine, but this is better. Airlines are susceptible to economic cycles and, depending on your seniority (or lack thereof), your job security could be at risk. There are adjustments that need to be made for family and friends, interests, and hobbies. You really don't have the luxury of being anywhere every Wednesday for a meeting or practice. You don't usually get to be home for every holiday. But, with all its characteristics and facets, a career as a pilot can offer a very interesting and different lifestyle. I have never taken off and climbed through an early morning low-cloud layer without thinking about the beauty and marvels I get to see. I have never once walked on a bridge across the Seine River in Paris without being grateful that I get to do this for a living. I intentionally make an effort to not take it for granted."

Andy also writes articles about getting started in aviation at http://fltops.com, and he maintains a blog at http://simcocreative.blogspot.com. Both sites will provide interesting and useful articles on being a pilot and traveling the world.

DO I HAVE WHAT IT TAKES TO BE A PILOT?

"There are no short-cuts to anyplace worth going, and that is certainly true of the aviation profession," says John Hale, a pilot who flies Boeing 757s and 767s for American Airlines to destinations as close as St. Louis and as distant as Frankfurt, Germany. "Determination,

Revisiting Kitty Hawk

Today, in the skies above Kitty Hawk, North Carolina, private planes circle the once famous field where the first airplane flight took place on December 17, 1903. The entire flight was a scant 120 feet, but it proved beyond a reasonable doubt that people possessed the ability to build a machine that was able to fly.

Near the site of the Wright brothers' first flight stands a museum housing some of the remnants of their illustrious achievements. The Wright brothers were actually bicycle makers. The intricate tools they used on their bicycles are displayed in the museum, showing that even the first pilots had a precision and a mechanical aptitude that has been one of the cornerstones of pilot success ever since.

dedication, focus, hard work, and a lot of roller coaster rides as a kid are suggested. The sometimes all-night flights and the immense responsibility can take its toll. The job itself is exacting, demanding, and, at times, dangerous, but the views are unmatched in any other profession of which I know."

Most pilots have strong mechanical skills and very decisive personalities. It is very important for them to be able to make decisions quickly and accurately, sometimes under a great deal of pressure. Pilots must be responsible individuals, because the safety of their passengers depends upon them. They must also be reliable. "Reliability is important," says Andrew Targosz. "Due to the nature of

To Be a Successful Pilot, You Should...

- have mechanical and technical aptitude
- be flexible in terms of your living location and schedule—you may be on call 24 hours a day
- be able to accept responsibility for the lives of others
- be calm under pressure
- be decisive and able to make quick, yet intelligent, decisions in times of danger or stress
- be willing to continue your education and training for your entire career

the job—having different schedules from week to week—you may have varying start times. A good pilot will always keep track of their schedule so they don't miss a flight. We have to be dependable in order to always be where we are scheduled."

Pilots also need to be flexible in terms of their living location and their schedules. Even pilots who work for large commercial airlines often spend nights away from home in different cities or countries. And with the mergers and closings of some of the large airlines, further flexibility will be needed in the future. Today's commercial pilot cannot expect to spend his or her entire life flying and building up seniority with a single carrier. "You can't carry seniority from airline to air-

line," says Andrew. "Say you were flying for an airline for many years—and that company goes out of business; if you get a job flying for another airline, you start at the bottom rung, no matter you're experience. Unfortunately, in this economy, this happens a lot."

Since competition for flying jobs is so fierce, pilots should be willing to move around from job to job. It may not be unusual for a pilot to work for a charter service as a flight instructor or for a private company as he or she accumulates flying hours and builds experience.

Besides certain personality characteristics, airline pilots are required by the FAA to meet certain physical requirements. They must be at least 23 years old, be in good health, and have 20/20 vision with or without glasses. They must have good hearing and no physical disabilities that could impede their performance. They must also pass periodic and random drug screens. "Good health is an important quality," says Andrew. "Pilots have to undergo three physicals a year—two with an FAA doctor, and one with your personal doctor. If we ever suffer a heart attack, or show signs of diabetes, we are automatically grounded."

Pilots must also continue to learn throughout their careers and will be expected to prove that they are still flight-ready. "We have to take a flight test every nine months—flight simulator tests and a question and answer portion," explains Andrew. "Simulator tests are four hours long. We also have ground school, where we learn evacuation and ditching procedures. Pilots need to study in order to pass these exams if they want to be able to keep flying."

The Air Line Pilots Association, International represents more than 52,000 pilots at 35 U.S. and Canadian airlines. A few airlines have their own pilot unions.

As with other unions, pilots pay scheduled dues to the union in exchange for a package of services that includes collective bargaining for wages and benefits.

HOW DO I BECOME A PILOT?
Education

"There are so many different ways you can become a pilot," says Andrew Targosz. "When I was in high school, I considered joining the military after graduation to become a pilot, but there were already too many pilots returning from the Vietnam War. I always wanted to fly military airplanes, but there were no training slots available. I then entered a flight school while I was in high school. By the time I graduated high school, I already had my pilot's license. I studied aviation in college, but continued to gather my ratings while I was working toward my degree. After I graduated from college, I started flying corporate planes, and then some commuter airlines. Then I completed some training to fly a 727 in Atlanta; at the same time I happened to interview for a New York company to fly DC9s. I flew these jets for about a year. I then interviewed with a major carrier and, based on my experience with the DC9s and 727s, was hired."

High School

Completing high school is a must if you are interested in becoming a pilot. You should take high school classes in mathematics, particularly algebra and geometry. Physics, shop classes, and meteorology are also helpful. Good activities in which to participate include sports that may improve hand-eye coordination, a local ham radio club, and flight-oriented organizations. As with driving, 16 is the age at which you can begin taking flying lessons.

Postsecondary Training

There are two main routes to gaining flight experience: military training and civilian training. Military pilot training is a two-year program for which a college degree is normally required. The first year is spent learning flight basics, including classroom and simulator instruction, as well as officer training. The second year is spent training in a specific type of aircraft. Following completion of this training, pilots are expected to serve at least four years before they can leave the military and pursue a civilian flying career. Approximately 50 percent of airline pilots receive their training via the military. John Hale trained to become a pilot in the U.S. Air Force. "Pilot training was challenging," he recalls. "It was a lot of hard work, but it was also exhilarating. On those cold, clear western Texas mornings during pilot training I would sometimes look down on the earth from my perch seated in the jet and think to myself, 'this must be what a hawk feels like.'"

Outside the military, there are approximately 600 flight schools certified by the FAA. The cost of flight training, however, is one of the major drawbacks to this approach. A flight education may cost upward of $10,000, once classroom and air time are paid for.

All airline pilots who are paid to transport passengers must have an airline transport pilot's license issued by the FAA. To obtain this license, pilots must be at least 23 years old, have accrued at least 1,500 hours of flying experience (including night and instrument flying), pass FAA written and flight examinations, and have one or more advanced equipment ratings.

According to the U.S. Department of Labor, "Initial training for airline pilots typically includes a week of company indoctrination; three to six weeks of ground school and simulator training; and 25 hours of initial operating experience, including a check-ride with an FAA aviation safety inspector."

Flying hours are placed in a logbook either by a flying instructor or by the pilots themselves. These logbooks serve as important records of the pilot-in-training's flying time. Besides accumulating flying hours, pilots must sometimes complete in-flight tests. These tests are called check-rides. During check-rides, the pilot's flying performance is rated by the flight instructor, and a pass or fail is given.

Quality, as well as quantity, of flying hours is important. Pilots need to fly in a variety of conditions and get experience in a variety of aircraft. In addition to time spent actually flying, pilots in training test their skills in flight simulators, which simulate such flying scenarios as night flying, thunderstorm flying, and landing without the use of an engine. High scores in flight simulators can translate to better job opportunities.

Certification or Licensing

Licensing of all pilots is governed by the FAA. To obtain a commercial pilot's license the pilot must have completed the designated number of hours, and must pass an in-flight test and a detailed written test.

Closely related to licensing is obtaining an instrument rating. Instrument ratings show that a pilot is able to fly based on reading instruments alone, without the help of visuals such as landmarks or clouds. These ratings change as pilots progress from flying single-engine to multiengine planes, all the way up through jets.

Additionally, airline pilots who complete rigorous training and background screening are deputized as federal law enforcement officers and are issued firearms. This training prepares them to protect the cockpit against hijackers and intruders, if necessary.

Internships and Volunteerships

You won't be able to participate in an internship or volunteer opportunity as a commercial pilot, of course, but you will be able to serve as an intern or volunteer and learn about aviation operations and the industry in general at an airline, aviation company, aviation association, or related organization. Southwest Airlines, for example, offers summer, spring, and fall internships in a variety of areas, including flight operations, dispatch, ground operations, public relations, and technology. College students typically participate in internships as part of their program. If not, college career services offices can help students locate rewarding opportunities.

WHO WILL HIRE ME?

About 79,000 pilots work for the major airlines. The remaining civilian pilots work for smaller airlines, charter services, private corporations, or the government.

Pilot jobs are heavily concentrated in states like California, Texas, Georgia, Washington, Nevada, Hawaii, and Alaska. These states have a higher flying activity relative to their population. More than one-third of all pilots work out of large centers like Dallas-Fort Worth, Los Angeles, San Francisco, New York, Chicago, Miami, and Atlanta. Seattle, Washington, D.C., Denver, and Boston are also large hubs in which many pilots are based.

It will most likely be impossible for the newly licensed pilot to secure a job with a large airline. Airlines require at least 1,000 hours of flying time, preferably in multiengine aircraft. The average new hire at regional airlines has more than 2,000 hours; the average new hire at the major airlines has almost 4,000 hours. Since most beginning pilots have not accumulated this many hours, they may want to start gaining experience by applying directly to charter companies. "In the airline industry," says Andrew Targosz, "it takes a lot of luck to get hired—everything is done by seniority. If you are able

Faster Than the Speed of Sound

In 1947 Bell Aircraft asked a 24-year-old Air Force captain named Chuck Yeager to attempt something no one had ever done before. Bell wanted the World War II flying ace to fly one of its planes, the X-I, faster than the speed of sound.

The X-I was a one-seater with small, razor-thin wings and was known for its difficult maneuverability. Nevertheless, Yeager flew nearly 700 miles per hour, creating what is known as a sonic boom, and launching the era of the modern jet aircraft.

In 1997, on the 50th anniversary of his remarkable and historic feat, Chuck Yeager took an F-15 Eagle up into the wild blue yonder, and once again, broke the speed of sound.

to apply for a job right when the airlines are hiring, then you'll be able to fly the bigger and better airplanes. The bigger airplane, the more weight, the more salary you can receive."

"My first job in aviation was with the flight school in Nashville where I was training," says Andy Simonds. "When I finished my flight instructor training (and became a certified flight instructor), I started teaching full time. I also started filling in on some of the charter flying the school was doing. After a fairly short time (18 months), I was hired at my first airline, American Eagle. I was based in Nashville (where I was living) and I had 12 days off per month. I first thought that we got 12 days off a year and that sounded pretty

Related Jobs

- agricultural pilots
- check pilots
- flight instructors
- flight-operations inspectors
- helicopter pilots
- navigators
- remotely piloted vehicle controllers
- test pilots

a seniority list (getting hired) at an early age is important for many facets of your career, particularly if it is at the airline where you want to spend the bulk of your career. Seniority determines what you fly (aircraft), which seat (captain or first officer), where you fly (destinations), monthly schedule, vacation schedule, pay schedule, and other important parts of everyday life."

Other employment possibilities for the beginning pilot include sightseeing companies, governmental agencies, private industry, and agricultural flying. Many of these smaller organizations require fewer hours of flying experience.

good, compared to flight instructing, so 12 days a month was like vacation. Even though the money was terrible, I loved the job. I knew what it paid before I took the gig, so there were no disappointments. It was a lot of flying in multiengine turbo-prop airplanes, and it was preparing me well for bigger aircraft and better jobs. It took me nine years from that point to get to American Airlines but, in the process, I worked for TWA, Airborne Express, and a second stint at American Eagle when I was furloughed from TWA (an occupational hazard)."

Andy says that it is important to get hired by an airline as early as possible in your career. "We choose our schedules based on seniority," he explains. "Seniority is determined by date of hire so anyone who was hired before I was (in 1998) is senior to me. In a seniority-driven system (which most airlines use), he or she who gets there first wins. So, getting on

WHERE CAN I GO FROM HERE?

Advancement for pilots is typically limited to other flying jobs. A pilot might start out as a flight instructor, accumulating flying hours while he or she teaches. With a bit more experience, he or she might fly charter planes or perhaps get a job with a small air transportation company. Eventually, he or she might get a job with an airline. Competition for these jobs, however, is fierce, and only the best and most qualified pilots are considered.

Once a pilot breaks into the airline industry, advancement hinges almost completely on seniority. The copilot may spend anywhere from five to 15 years before becoming captain. If a pilot moves from one airline to another, he or she must start over again at the bottom and rebuild seniority.

WHAT ARE THE SALARY RANGES?

There are several factors that influence a pilot's salary level, including seniority, type of aircraft flown, experience level, and the airline he or she works for. Airline pilots also earn more for international and nighttime flights. The U.S. Department of Labor reports that mean annual earnings of airline pilots and copilots employed in scheduled air transportation were $111,680 in 2008. The lowest 10 percent earned less than $55,330 while the highest paid 10 percent earned $166,400 or more. Very experienced pilots at large commercial airlines may make $200,000 or more annually.

Most pilots receive a benefits package that includes life and health insurance, retirement plans, and disability payments. Also, for the large commercial airlines, travel benefits are usually included in the employment package; pilots and their immediate families fly free on most airlines.

WHAT IS THE JOB OUTLOOK?

Employment of pilots is expected to grow about as fast as the average for all careers through 2016, according to the U.S. Department of Labor. Although growth is expected in both airline passenger and cargo traffic—which normally would create more jobs—that growth will be offset by several factors. A weak economy is causing business and casual travelers to scale back the number of trips they take. Employment of pilots rises and falls

> ## Did You Know?
>
> There are 517 commercial service airports in the United States, according to the Air Transport Association.

with the condition of the overall economy. During recessions, when there is a reduced demand for air travel, airlines cut back on their flights and consequently on their employees. Commercial and corporate flying, flight instruction, and testing of new aircraft also decline during recessions, causing a decreased need for pilots in those areas. Another factor is the trend toward using larger aircraft, which allows more passengers and more cargo per flight and ultimately reduces the number of flights flown.

Because of the travel benefits and prestige, pilots tend to remain in the field once they have obtained a position. However, pilots who reach the mandatory retirement age of 65 will leave the industry, generating several thousand job openings yearly.

If you are considering a career as a pilot, you should be aware that you will face keen competition for jobs. Competition for pilots' jobs has intensified in recent years because of an increase in the number of qualified, unemployed pilots. During the recent restructuring of the airline industry, many pilots lost their jobs. Employment opportunities will be best at regional airlines and low-cost carriers—although Andrew Targosz says that there are downsides to working for these types of employers. "The major airlines are outsourcing most of their domestic flights to regional airlines—regional jets with 50- to 70-passenger planes," he explains. "Regional jet (RJ) crews and mechanics are typically paid less. Pilots who fly RJs will probably never see any international flying."

It will be difficult to land a job at one of the major airlines. "The way the economy is right now, the major airlines are furloughing," says Andrew. "This means that pilots don't necessarily lose their jobs—they have callback rights. It might be five to seven years before they call you back on the job. But if you want to fly, and have a love for flying, then there will always be a job for you. However, if you are going into this field for the promise of money, it's not there—at least for now. We are hoping that the economy will rebound, and the industry outlook will change. Right now, the way the major airlines are contracting and the regional jets are expanding, there isn't a lot of money to make in this industry. However, if things change, who knows?"

Tour Guides

SUMMARY		
Definition Tour guides lead groups of people on visits to sites of interest. Some guides lead short excursions that last only a few hours or a day. Other guides, sometimes called tour managers, lead groups of travelers on extended trips that can last anywhere from a few days to a month.	**Alternative Job Titles** Tour directors Tour managers **Salary Range** $15,470 to $23,270 to $100,000+ **Educational Requirements** High school diploma **Certification or Licensing** Voluntary **Employment Outlook** Much faster than the average	**High School Subjects** Foreign language Geography History Sociology Speech **Personal Interests** Entertaining/performing Helping people: personal service Teaching Travel

"A tour guide is someone who is an expert, a storyteller, a leader, and a problem solver," says Kevin Doran, the owner of Two Foot Tours. "A guide knows all about the local culture and neighborhoods, history, geography, points of interest, attractions, museums and architecture, shopping, entertainment, and restaurants. A tour guide must also direct the driver, interface with the clients, resolve emergencies, know where buses may legally drive and safely park for passengers to get on and off, coordinate admissions, make lunch recommendations, keep to a schedule, and know the location of a sanitary bathroom. To make this more challenging, guides are often expected to do this while facing backward on a moving bus all the while creating an imaginative route with entertaining commentary. Professional guides also respect ethical standards pertaining to employers, intellectual property, and the tourism community. Professional guides are not made overnight."

WHAT DOES A TOUR GUIDE DO?

Imagine traveling somewhere completely unfamiliar to you—a foreign country or a different city. You would probably have many questions: Where are the best places to eat? How do you get to your hotel? What sights should you see? If you were traveling abroad, you might also have questions about language, customs, and the value of your money in foreign currency. Now imagine being the person who has the answers to all those questions, and you will have an idea of what it is like to be a *tour guide.*

Tour guides escort groups of people who are traveling to different cities and countries. Essentially, their job is to make sure that their travelers have a safe and enjoyable trip by planning and overseeing every detail of the tour. Some guides take passengers on short excursions, which

may last a few hours, a full day, or even overnight. For example, travelers visiting Los Angeles might take an all-day tour of Beverly Hills and Hollywood. Other guides—who are sometimes also called *tour managers*—accompany their groups on longer trips, lasting anywhere from a few days to a month. These longer trips generally involve travel to foreign locations and may include visits to several different cities or countries.

Tour guides who own their own businesses are responsible for making all the necessary arrangements for a trip prior to departure. Depending upon the length and type of the trip, this could involve several different things. They might book airline flights, ground transportation such as buses or vans, hotel rooms, and tables at restaurants. If anyone in the tour group has special needs, such as dietary requirements or wheelchair accessibility, the guide must attend to these needs in advance. Guides also plan the group's entertainment and make any necessary advance reservations. They may reserve tickets to plays, sporting events, or concerts. They may also contact other guides with specialized knowledge to give group tours of various locations. For example, for a group visiting Paris, the tour manager might arrange for a guided tour of the Louvre one day, and another guided tour of the famous Left Bank on another day.

Tour guides who work as freelancers for tour companies do not usually do this type of advance booking. The tour company handles all booking, and guides are tasked with confirming the reservations.

To Be a Successful Tour Guide, You Should...

- like people and enjoy communicating with them

- be able to cope with emergencies and unplanned events

- be in good health and physical condition

- feel comfortable being in charge of large groups of people

- have a good sense of humor and be fun-loving

- be highly organized

Once the plans have been made and the tour begins, the guides' duties may include almost anything that makes the trip run smoothly. They must make sure that everything goes as planned, from transportation to accommodations to entertainment. They must see to it that passengers' luggage is loaded and routed to the proper place. They must either speak the local language or hire an interpreter, and they must be familiar enough with local customs and laws to ensure that no one in the group unwittingly does anything illegal or offensive. They must make sure that all members of the group stay together so no one gets lost, and that they are on time for various arrivals and departures.

In addition to merely monitoring the particulars of the trip and chaperoning group members, the tour guide is responsible for educating the group about the places they are visiting. Therefore, guides are generally very familiar with the locations they are visiting and are able to answer questions and provide a sort of entertaining and educational commentary throughout the trip.

Guides must be prepared to deal with unexpected difficulties or changes in plans. If a point of interest is closed, if a hotel has failed to reserve enough rooms, or if weather conditions delay travel, it is up to the guide to make alternative arrangements. The guide is also responsible for attending to whatever needs travelers might have. This could include anything from calling the hotel concierge for extra blankets to taking a sick traveler to the hospital. While tour group members are traveling in unfamiliar territory, they depend upon the tour guide for almost everything.

WHAT IS IT LIKE TO BE A TOUR GUIDE?

Kevin Doran is the owner of Two Foot Tours (http://www.twofoottours.com), a private tour company in south Florida. He is also the president of the Professional Tour Guide Association of Florida. He has worked in the field for six years. Before that he was employed in the film and television industry. "The industry was changing rapidly," he recalls, "and when the company I worked for downsized from 300 employees to 50 I found myself without a seat when the music stopped. At 52 I awoke no longer trusting the security of employment. I looked for a business that suited my skills and interests that would liberate me from the structured office routine that often accompanies office jobs. I also wanted something that didn't require extensive start-up resources.

"I started Two Foot Tours providing private, customized tours of south Florida," he continues. "Miami is a great tourist city with a year-round tourism industry. Not only is the weather ideal but in many ways the city is maturing for the first time and is recognized as a cosmopolitan, multicultural city, with a growing art scene and international business, architecture, arts, and entertainment to accompany tropical beaches. I thought that it would be an ideal place to develop cultural tours to explain the fascinating culture to visitors from around the world.

"A normal day in my work life usually begins the night before with a review of the next day. My office is at home so I don't have to go too far to get to work. I feel most creative in the morning, so I like to write or do accounting in the morning and try to begin at 7 A.M. Morning tours usually begin around 10 A.M. A short tour will last about two to three hours, plus travel time. Sometimes I will do two tours in one day. It takes a lot of energy to talk for five to six hours, and I'm tired afterward if I do. On these days I don't do office work or have meetings later in the day unless it is necessary. On days without tours office work includes correspondence, writing bids and proposals, accounting, and research. I find the office stimulating and love working at home. It is much more productive than working in an office with coworkers because there is less interruption. I have the comfort and efficiency of food, music, and rest close at hand. I also have a gym and a pool and can schedule my time." Kevin says that it is easy, though, to become isolated when working at home. "Personal interaction is important in business psychologically, creatively, and practically," he says. "It sometimes requires effort to get out of the home/office, but it is absolutely critical in my opinion."

Kevin says that he wears many hats as a small business owner. "I create the tours so I spend considerable time researching and writing," he explains. "This begins with the morning paper every day, Internet, books, lectures, interviews, discussions, tours, and occasional classes. I enjoy this part of the business tremendously and spend about 30 percent of my time engaged in some form of content creation.

"I market the business," he continues, "so I'm involved in creating the brand identity and developing strategies to position and sell my services. As a small business I often mix marketing with selling. By this I mean that much of the branding and advertising of my services is done through public speaking, volunteer work with professional and community organizations, networking, and other activities. There is strategic thinking about how these efforts help me achieve my objectives, but pure marketing is something that I don't dedicate enough attention to. I spend 30 percent of my time involved in this part of the business and am considering professional help in this area.

"I schedule and lead tours so I must make the deals with vendors, arrange the venues, meet the tourists, and speak for several hours on tour days. I do bids and proposals and re-do bids and proposals. Less than half of the bids produce business. I wish there were a way to be more selective about which bids to focus on, but sometimes good jobs come from unexpected places and promising leads often lead nowhere. About 25 percent of my time is spent on this part of the tour business.

"I also do the accounting. I have a very simple accounting system. Other requirements concerned with legal requirements, licenses, and compliance for a tour guide service are minimal, so I probably only dedicate 10 percent of my time to this area."

Keri Belisle is a tour guide in California. She has worked in the field since 1990. "I happened upon the tour business while in college," she says, "and worked in that field until I graduated. My major was graphic art, but I found it too stressful, so I decided to continue on in tourism. I never went to an official tourism school or took any kind of tourism class. In 1990 I was hired by Contiki Holidays; they had a three-week intensive training course, which really helped prepare me. I love leading tours. I'm in my element. I enjoy imparting my knowledge and sharing sights and experiences with my passengers."

Keri works for several different tour companies. "One of the tour companies does day tours, and then I do longer tours (six to 14 days) for a few other companies," she explains. "On day trips, I show up to the office and get the supplies, itinerary, manifest, etc. The bus picks me up there. The day before, I research about where I'm going and what I'm seeing, whether it is a museum, national park, festival of roses, spa, or other destination. I always try to get fun and useful trivia. On overnight tours, preparation usually starts a week ahead of time. I gather my notes, books, music, and any thematic items. I tend to buy 'extras' for a tour that companies don't provide—prizes for bus games, birthday and anniversary presents, items to pass around the bus, etc. I usually fly to my starting point. My main job is to get people from point A to B, providing history, current events, geography, and humor along the way. I also am responsible for hotel check-in, meals, and included extras, among other tasks. My secondary job is to make sure everyone is happy and having a good time (really the primary job).

"I mostly lead tours to the Grand Canyon, Bryce Canyon, Zion, Joshua Tree, and Yosemite National Parks," she continues. "We also go to some of the missions in California. When I lead tours for the local tour company out of San Diego, some of the destinations we go to include The Getty Center, The Getty Villa, The Huntington Library, Norton Simon Museum, Museum of Tolerance, Los Angeles County Museum of Art, The Bowers Museum of Cultural Art, The Nethercutt Collection, The Ronald W. Reagan Presidential Library, The Griffith Observatory, The Los Angeles Times tour, the Federal Bank tour, Hearst Castle, and musicals (such as the *Phantom of the Opera*, *Cats*, *The Lion King*, and *Wicked*)."

Keri says that there are many plusses to working as a tour guide, including the chance to travel, flexibility regarding schedule, and the opportunity to be your own boss, meet new people, face different challenges daily, and continue to learn. She says that one of the main drawbacks to this career is that "many companies do not hire full time, so it is a juggling act to fill up your calendar, only to have a tour cancel, and then not be able to find something with which to replace it. Additionally, the days can be very long, and I never get to sleep in. And you have to pay for your own health insurance."

DO I HAVE WHAT IT TAKES TO BE A TOUR GUIDE?

Keri Belisle says that the most important personal and professional qualities for tour guides are "patience (and more

....
Pros and Cons
● ● ● ● ● ● ● ● ●

Kevin Doran details what he likes and dislikes about being a tour guide:

The things I like most about being a tour guide are that you can constantly improve your skills, there is no end to learning, and every tour brings new faces. It is wonderful to be outside, moving around, immersed in a world of beauty and ideas, and sharing knowledge with eager, happy people.

Every profession has its problems. In tourism the weather can affect business negatively, tour guides don't get rich, customers can be difficult, and like any job it can become routine if you repeat the same activity too often.
....

patience!), quick thinking, flexibility, organization, a sense of humor, a pleasant personality, being knowledgeable, and the ability to project authority. There are so many aspects to the job: getting along with others, keeping everyone happy, giving clear and concise instructions, giving interesting narration, accounting, keeping things on schedule, and common sense."

To be a good tour guide, it is imperative that you genuinely enjoy talking in front of a group. If you get nervous talking to large groups or don't enjoy being with groups of people, this is probably the wrong profession for you. Because you sometimes encounter tour group members who are difficult or demanding, it is also important that you be a patient person who likes people.

Tour guides should also enjoy having fun and helping their group members enjoy themselves. Another important trait is the ability to deal with unforeseen difficulties. If you hit traffic and can't be where you're supposed to be at 3:00, do you go to pieces or do you improvise? If someone gets sick on the tour, do you freak out, or do you deal with it? If you fall apart when things don't work out perfectly, this probably isn't the career for you.

Leadership and a take-charge attitude are also necessary in this job, where guides are relied upon to answer questions, deal with problems, and generally take care of travelers' needs. "There are so many rewards and challenges on a daily basis," Keri says. "Just when you think you know it all, something happens to change that. I can say that I actually like the challenges that I am faced with. I am always thinking on my feet."

Finally, tour guides need to be willing and able to work long hours. During a tour, guides are never really off duty, and this can mean phone calls in the middle of the night to resolve any problem a guest is having. For a tour manager, a workweek of 85 hours is not uncommon.

HOW DO I BECOME A TOUR GUIDE?
Education
High School

A high school diploma is the minimum educational requirement for becoming a tour guide, although many guides do have some postsecondary training. If you hope to become a tour guide, there are several high school courses you can take that will prepare you for the position and improve your chances of finding a job. Perhaps the single most valuable class is a foreign language. Tour guides who can speak a second language fluently will be in the greatest demand.

A good tour guide should have a grasp of his or her destination's history and culture; therefore, classes in social studies, sociology, geography, and history are excellent choices. Since knowledge of the arts is also important on many tours, courses in art history or appreciation would also be helpful. Because much of the tour guide's work is in communicating with people—and often may involve speaking to groups of travelers—classes in speech can help prepare you for this job.

In addition to taking the right preparatory classes, you may be able to gain experience finding a part-time or summer job working as a tour guide. Local historical sites or museums often use part-time workers or volunteers from the community to conduct tours.

Postsecondary Training

Although there is no formal educational requirement for becoming a tour guide, many guides do have some postsecondary training. Many two- and four-year colleges offer courses in tour management and tour guiding. Some trade and professional schools also offer tour guiding and tour management programs, and a few large travel agencies offer classes to teach employees how to conduct tours.

Some tour guides—especially those interested in leading special inter-

est tours—have bachelor's or master's degrees in various subjects. For example, someone with a degree in architecture might lead a group of travelers through Italy's churches. Or a guide with a degree in American history might lead a group on a tour of Civil War battlefields. If you hope to combine your interest in a particular field with a career as a tour guide, you should focus heavily on your area of

An Interesting Experience

"I had an experience a year ago that still makes me smile," says tour guide Kevin Doran. "I do tours for the Miami Council for International Visitors and was asked to do a private tour for a teacher from Tibet who was studying women's education as a guest of the U.S. State Department. We had a translator and a driver, so there were four of us on the tour. I was prepared to give her a city tour and explain the history and culture of Miami, but when we got to Miami Beach and walked down to the ocean she saw the ocean for the first time in her life and the effect was overwhelming. She had lived her entire life above 12,000 feet on the Tibetan plateau and had never seen anything like it before. I must admit that it was a spectacularly beautiful day and the water was blue and sparkling. She took off her shoes and walked in the water and asked if she could spend the rest of the tour right there. She collected shells and had the experience of a lifetime in the simplest of pleasures. It was the easiest tour I've ever done. I will never forget what I learned that day."

interest while taking supplemental classes in public speaking and travel and tourism, where available.

Certification or Licensing

The National Tour Association offers the voluntary certified tour professional designation to candidates who meet education, employment, and service requirements; complete coursework; and complete a learning portfolio.

Internships and Volunteerships

You can get experience working as a tour guide by volunteering with a local company or participating in an internship. Your high school or college career services office may be able to provide some suggestions on where to find opportunities.

WHO WILL HIRE ME?

There are approximately 40,000 tour and travel guides employed in the United States. The major employers of tour guides are, naturally, tour companies. However, most of the major tour operators prefer experienced guides and are unlikely to take a chance on a beginner. Therefore, it may be wise to start on a smaller level. If you are familiar with a certain city, region, or local attraction, you might want to apply for a job giving tours of that area or site. Industrial plants, colleges and universities, chambers of commerce, museums, historic sites, zoos, and parks may all hire guides to give short, informational tours of their facilities. Once you have gained some experience in touring, you might then graduate to longer range,

over-the-road tours conducted by tour companies.

To obtain a list of tour companies, you might approach a local travel agency and ask for help; they are likely to have a trade publication that lists all the major companies in the travel and tourism industry. The reference desk at your local library may also be able to help you compile such a list. Remember that the fastest growth in this field is likely to occur in inbound tourism; therefore, the majority of job openings will be in large cities and areas of heavy tourist traffic, such as Disney World or New York City. In your job search, be sure to thoroughly explore these high-tourist areas. Check with travel agencies in those cities or consult yellow page listings under "tours" or "tour operators." Once you have a list of tour companies, you might send resumes and cover letters to the ones that interest you.

WHERE CAN I GO FROM HERE?

Most guides begin their careers working part time on one-day tours. As they gain experience, they start to travel with experienced guides on a particular tour until they have mastered the itinerary and the necessary information and are able to lead that tour on their own. Career advancement can take the form of leading more complicated tours, or of specializing in a certain type of tour—such as tours to specific destinations or tours that focus on a particular interest. Guides who are good at their work often build up a following of repeat customers who sign up for their tours. These popular guides may then be able to move to a higher-paying tour company or, with the right combination of business skill and investment capital, open their own agencies.

Some tour guides become *travel writers*, reporting on various destinations for the many travel-oriented magazines and newspapers. Others may move into the corporate world, planning travel arrangements for company business travelers.

WHAT ARE THE SALARY RANGES?

The work of a tour guide is often seasonal—extremely busy during the peak travel times of May through October, and much slower in the off-season. Annual salaries for tour guides ranged from less than $15,470 to $38,280 or more in 2008, according to the U.S. Department of Labor. The median salary for tour guides was $23,270. Tour company owners can earn more than $100,000 annually. While traveling, guides receive their meals and accommodations free, as well as a daily stipend to cover additional expenses.

Depending upon what tour operator they work for, tour guides may also receive a benefits package that includes sick and vacation time, health insurance, and profit sharing. Guides often also receive discounts from hotels, airlines, and transportation companies.

WHAT IS THE JOB OUTLOOK?

The travel and tourism industry is expanding and is expected to continue its growth through 2016, although the recent

Hot Domestic Travel Destinations, 2009

Travel agents rate the following U.S. destinations as the most popular for tourists:

1. Orlando, Florida
2. Las Vegas, Nevada
3. San Francisco, California
4. Los Angeles, California
5. Miami, Florida
6. Honolulu, Hawaii
7. San Diego, California
8. Washington, D.C.
9. Chicago, Illinois
10. New York, New York

Source: American Society of Travel Agents

economic downswing has dampened this growth to some extent. The market for package tours and special-interest tours, such as nature tours, wildlife tours, or architectural tours, is growing. This should create the need for more of these types of tours, and therefore, more tour guides.

Another form of tourism that is on the upswing is inbound tourism—guiding foreign visitors through famous American tourist sites. To many foreign travelers, America is a dream destination, with tourist spots such as Hollywood, New York, Disney World, Yellowstone, and other nature-, history-, or culture-rich areas drawing millions of foreign visitors each year. Inbound tour guides with both cultural and foreign-language skills, especially Russian, German, and Japanese, would do the best in this growing subfield. Job opportunities in inbound tourism will likely be more plentiful than those guiding Americans in foreign locations. The best opportunities in inbound tourism are in large cities with international airports and in areas with a large amount of tourist traffic.

Even though tourism is on the rise, prospective tour guides should realize that there is intense competition in this field. Tour guide jobs, because of the obvious benefits, are highly sought after, so the beginning job seeker usually finds it difficult to break into the business. It is also important to remember that the travel and tourism industry is affected by the overall economy. When the economy is depressed, people have less money to spend and, therefore, travel less. They may take shorter tours instead of multi-day trips, which will reduce revenue for tour guides.

Travel Agents

SUMMARY

Definition
Travel agents help clients make arrangements for both business and leisure travel. They make reservations for air travel, car rental, hotel accommodations, cruises, and packaged tours. To obtain fares, schedules, and availability, travel agents consult a variety of sources, such as online computer reservation systems, guidebooks, and other published materials.

Alternative Job Titles
None

Salary Range
$18,770 to $30,570 to $65,000+

Educational Requirements
High school diploma; some postsecondary training recommended

Certification or Licensing
Recommended

Employment Outlook
Little or no growth

High School Subjects
Business
Computer science
Geography/social studies

Personal Interests
Computers
Helping people: personal service
Travel

"As a travel agent," says Heather Dolstra, "I have been all over the world and have memories that will last a lifetime. You will work hard and you probably won't get rich (unless you come up with a better Web-based mouse trap), but you will never be bored and you will never stop learning. And the skills are portable. If you need to move to another city or state, you can pick up and go immediately. Many travel agents are now home based or working remotely, which means there are many flexible working arrangements possible."

WHAT DOES A TRAVEL AGENT DO?
Travel agents help clients make travel plans both for business and for pleasure. They serve as salespeople and consultants to their clients, providing them with information, guiding them through a decision-making process, and selling them the travel product that they decide upon. Agents may make reservations for their clients for air travel, hotel accommodations, rental cars, cruises, package tours, and rail travel.

Agents first determine their clients' needs, interests, time constraints, and budgets. Arranging leisure travel may be quite different from arranging business travel. Clients planning a vacation may want to spend considerable time in planning, learning about their different options, and looking at brochures and travel videos and DVDs. Business travelers, on the other hand, often have very specific, already-established requirements for travel and want the arrangements made quickly. Whichever the type of travel, agents work with the client to design a trip that meets his or her specifications. They consult a variety of sources, such as published materials and online computer reservation systems, for airline departure and arrival times, fares and availability, hotel and car rental rates, and cruise and tour packages. They present information to the client and offer a choice of travel plans in an easy-to-understand fashion.

Once the client has made a decision, the agent must make the necessary reservations, issue electronic tickets or confirmation numbers, and, in some cases, collect payment. Making reservations may be done via an online reservation system or via telephone. Most bookings for airlines, hotels, and rental car companies are made electronically through a computer reservation system. To book electronically, the agent enters the client's information into the computer system, which sends it to the appropriate travel supplier. The supplier—whether it is an airline, hotel, or car rental company—reserves a seat, a room, or a car for that client. In the case of air travel reservations, the travel agent prints out an actual airline ticket on an on-site ticket printer. For hotel and car reservations, the agent receives only a confirmation number, which is printed out and passed on to the client. Tour packages and cruises are typically reserved over the telephone rather than electronically.

Travel agents may also serve as travel consultants. They explain about customs regulations, passports and visas, health permits, and foreign currency exchange rates. They may advise travelers on what kind of clothing to pack, baggage and accident insurance, traveler's checks, sightseeing, and restaurants.

Agents often sell package tours that are developed by another organization. In some cases, however, a group of people may ask an agent to design a tour for them. In this case, the agent sets the itinerary and makes all the necessary reservations for transportation, accommodations, meals, and activities. In addition, the agent might be responsible for providing a tour guide and publicizing the tour through brochures and advertisements.

Agents serve as bookkeepers to handle the complex details of all the trips they schedule. They serve as the go-between for the client and the supplier, making sure that arrangements are made and understood properly. They provide detailed itineraries, confirmation numbers, and tickets to their clients. In many cases, such as with air travel and cruises, they obtain payment from the client and route it to the appropriate vendor.

Travel agents may also promote their services by giving talks at social, community, or club meetings, or by suggesting company-sponsored trips to business groups. In a small agency, one or two agents may do all the sales, bookkeeping, promotion, and clerical work. In a larger agency, however, there may be clerks and secretaries who help with the clerical aspects of the job, such as filing, answering phones, and opening mail.

Although many travel agents deal with all kinds of clients and all kinds of travel, in some cases they may specialize in either a particular kind of travel or a particular destination. Some agents may deal primarily in business travel, for example. Others may work principally with group tours or with particular geographic destinations such as Europe, Asia, or the Middle East.

WHAT IS IT LIKE TO BE A TRAVEL AGENT?

"I entered the travel industry as a fluke," says Joanne Gardner, the owner of the THE Travel Specialist, a travel agency in Wheaton, Illinois. "I spent my college breaks traveling in Europe because I loved it. I graduated with a degree in social welfare and was working as a social worker, then the funding for the program was cut and I was jobless. My aunt told me about a travel agency that was hiring and I got the job. I loved it from the start and never looked back. It has been 36 years now.

"The pros of the work," she continues, "include traveling and helping others

Lingo to Learn

Airlines Reporting Corporation (ARC) An autonomous corporation created by the domestic airlines that appoints travel agencies to sell airline tickets and oversees the financial details of tracking payments to airlines and commissions to agencies.

booking A reservation.

coach The economy class on an airline.

computerized reservation system (CRS) Any of several computer systems allowing immediate access to fares, schedules, and availability, and offering the capability of making reservations and generating tickets. The two most commonly used are Sabre and Apollo.

confirmation number An alphanumeric code used to identify and document the confirmation of a booking.

fam Abbreviation for "familiarization" trip or tour. A low-cost or free trip or tour offered to travel agents by a supplier or group of suppliers to familiarize the agents with their destination and services.

layover A stop on a trip, usually associated with a change of planes or other transportation.

luxury class The most expensive accommodation or fare category.

travel. Making people see the benefits of just what they need to make their vacation a treasured memory is very fulfilling. Getting to travel myself just makes it better. The cons of the job include the fact that you have no control over what happens once you make your sale. You turn the trip over to the suppliers and hope they come through for the client. These days, too, travel agents are being second-guessed by their clients, who think they can do it better and cheaper themselves. The Internet has given them that impression. However, they need to remember that good agents have great contacts and have 'been there and done that' so they can provide personal recommendations specific to the client. Suppliers who want to charge us for selling their products are also a frustration. Everyone wants us to work for nothing and that is not the way a business survives. Travel industry people are a tough breed. We have had to reinvent our lives over and over to survive, and those of us still standing are very happy to be here. We love the industry and the people in it."

Heather Dolstra is the owner of Democracy Travel Inc., in Washington, D.C. She has worked in the field for more than 30 years. "I grew up in Europe as a Department of Defense brat and had the opportunity to travel from a very early age," she recalls. "After getting both a BA and MA, I still had no idea what I wanted to do as a grown-up. I did a little legislative work for a political group here in Washington, D.C., but nothing seemed 'right.' One afternoon I accompanied my mother to her travel agent's office, where they were planning a trip through Italy for my parents. I started flipping through the *Official Airline Guide* (this was 1978) and the agency owner suggested I might want to consider a travel school. Long story short: I did an intensive summer program and got hired out of the class by a fellow student and owner

of a new travel agency. I did fall into this career, but I was prepped by my early life experiences and was always drawn to anything having to do with travel."

As the owner of her own business, Heather must handle a variety of duties. "For better or worse," she says, "I handle both reservations and management issues. That means that I may be attending to questions from the bookkeeper in between client calls. I also deal with both leisure and corporate travel. I personally like the variety of working with both business clients and vacationers. Working on vacations (family travel, honeymoons, spring break, summer holidays) requires a lot of emotional energy. Trying to match a client to the best product or opportunity is not unlike being a therapist, and it requires diligence, tact, and time. Having a few straightforward corporate requests can provide a little breather. Every day has deadlines: ticketing, hotel, deposits, promotional dates ending. Clients also have their internal deadlines of when they want to have their business concluded. Nothing like a little pressure! And in between all of this client-directed business, we also have to read supplier material and keep up with new properties, new tour operators, and current events worldwide."

Just like any career, Heather says that there are both drawbacks and rewards to working as a travel agent. "The cons are that there is a definite level of stress involved in trying to meet client expectations 24/7," she explains. "You cannot walk out of the office when there are deadlines. And there are always deadlines. Competition from the airlines and from online suppliers is demoralizing. Clients sometimes do not seem to value what you have to offer, and that can be discouraging. The upside to working in this field is the opportunity to travel paired with the need to travel for professional development. In addition, there is the vicarious pleasure from working out a trip for clients and hearing about their experiences. Dealing with clients who come back again and again means that you increase your circle of friends as well. You go through all of life's stages with your best clients! For people who love learning and love people, this is an ideal job."

Kristle Corcuera has been working as a travel agent for ProTravel Network since 2008. "I was interested in becoming a travel agent because of all of the travel perks that come with the job," she says. "I get great discounts on travel and I am often invited to check out new resorts/hotels for free to recommend to my clients. My primary duties are to research prices, suggest vacation ideas, and most of all to fulfill the client's wants/needs for their vacations, flights, car, hotels, etc. I also give presentations about certain locations to help promote business to new getaways. Other job duties include building your clientele and constant client follow-up. Sending emails, passing out business cards, and word of mouth are key to building your business."

DO I HAVE WHAT IT TAKES TO BE A TRAVEL AGENT?

The key requirement for being a good travel agent is the ability to communicate.

To Be a Successful Travel Agent, You Should...

- have excellent communication skills
- be enthusiastic, patient, and courteous in dealing with people
- have some sales ability
- have an interest in travel
- be precise and detail-oriented in your work habits
- be able to concentrate on your work despite interruptions and a lack of privacy
- be proficient using computers and the Internet

If an agent doesn't communicate well with the client, the client may not trust that agent to make the right arrangements and get the best deal. "Customer service is a big part of the job," says Kristle Corcuera. "Having great social skills and being attentive is very important. You need to make sure to listen to your clients in order to give them the greatest experience. The more they enjoy their vacation, the more they will book with you and also recommend you to their friends."

Some selling skills are also important. A travel agent doesn't have to use high-powered sales tactics, but they should have some assertiveness in order to help customers make a timely decision regarding their travel plans. "Sales skills are a must, as is good knowledge of geography," says Joanne Gardner. "You need to know what direction you are sending people! You are interacting with people at all times—clients and suppliers—and need those skills. Anyone can be trained in the technology of the business, but you need to come with personal skills."

Patience and a tolerance for stress are also necessary to success in this field. During peak travel times, such as holidays and spring break, the travel agent's job can become very stressful.

"The industry is a great business to be in," says Joanne, "albeit a constantly changing one. You need to be flexible and able to handle changes in your job. Talk to someone in the position you think you would like and get inside information as to how they actually perform their job. Take the appropriate classes in college, be it travel-specific classes or marketing."

HOW DO I BECOME A TRAVEL AGENT?

Education

High School

A high school diploma is the minimum requirement for becoming a travel agent. If you are interested in pursuing a career as an agent, be certain to include some computer courses, as well as typing or keyboarding courses, in your class schedule. Since much of your work as a travel agent will involve computerized reservation systems, it is important to have basic keyboarding skills and to be comfortable working with computers and the Internet.

Because being able to communicate clearly with clients is central to this job, any high school course that enhances communication skills, such as English or speech, is a good choice. Proficiency in a foreign language, while not a requirement, might be helpful in many cases, such as when you are working with international travelers. Finally, geography, social studies, and business mathematics are classes that may also help prepare you for various aspects of the travel agent's work.

You can also begin learning about being a travel agent while still in high school by getting a summer or part-time job in travel and tourism. Interested high school students should try to find a job in a travel agency. If finding a part-time or summer job in a travel agency proves impossible, you might consider looking for a job as a reservation agent for an airline, rental car agency, or hotel.

Postsecondary Training

Currently, most travel agencies do not require their agents to have college degrees. Increasingly, however, travel agencies are seeking applicants with college degrees, and when it comes to advancement, the agent with more education is likely to have an edge over those with less. Some colleges offer two- and four-year degrees in travel and tourism. If your college of choice does not offer a specific degree in travel and tourism, a degree in geography, communications, computer science, history, business, or a foreign language might be equally helpful. Other good college courses to take include computer science, world history, and accounting. The American Society of

Travel Agents (ASTA) provides a list of member schools at its Web site (http://www.asta.org).

Another option for prospective travel agents is to take a short-term course in travel specifically designed to prepare you for work in this field. Such courses are typically between six and 18 weeks in length and are offered by community colleges, vocational schools, and adult education programs. The American Society of Travel Agents offers virtual seminars on industry-related topics, and there are a number of travel schools that combine home study with on-site training to prepare future agents. The Travel Institute is also a major provider of educational programs for travel professionals. It offers travel agents a number of other programs, such as sales skills development courses and destination specialist courses, which provide a detailed knowledge of various geographic regions of the world.

Certification or Licensing

Travel agents may choose to become certified by The Travel Institute. The institute offers the certified travel associate and certified travel counselor designations to applicants who complete education and experience requirements. While not a requirement, certification by the institute will help you advance your career. Additionally, the National Business Travel Association offers certification to travel agents.

Most states do not require travel agents to be licensed or registered. However, there are exceptions, so it is important

---

Related Jobs

- airline reservation agents
- concierges
- crating-and-moving estimators
- rental car agents
- sales representatives (shipping services)
- tour guides
- tour operators
- traffic agents

---

to check the requirements for the state in which you will be working.

To be able to sell passage on various types of transportation, you must be approved by the conferences of carriers involved. These are the Airlines Reporting Corporation, the International Air Transport Association, and Cruise Lines International Association. To sell tickets for these individual conferences, you must be clearly established in the travel business and have a good personal and business background. Not all travel agents are authorized to sell passage by all of the above conferences. Naturally, if you wish to sell the widest range of services, you should seek affiliation with all three.

Internships and Volunteerships

Most college travel and tourism programs will require that you participate in an internship program with a travel agency or another company that employs travel agents. Schools often have partnerships set up with local travel agencies or corporations that have travel planning departments. Participating in an internship will provide you with an opportunity to make valuable contacts and learn more about the field from experienced travel agents. Internships can last anywhere from four months to a year.

WHO WILL HIRE ME?

There are approximately 101,000 travel agents currently working in the United States. Agents may work for commercial travel agents, in the corporate travel department of a large company, or be self-employed. Sixty-seven percent of agents work for travel agencies; approximately 13 percent are self-employed.

Most new travel agents find jobs by applying directly to travel agencies. The job seeker might pinpoint a geographic area he or she is interested in and send resumes and cover letters to all the agencies in that area. Job openings for travel agents are also sometimes listed in the classified section of local newspapers, travel magazines, and industry publications. Industry publications often include a classified section that lists job opportunities for agents. Jobs are also listed on the Web sites of professional associations such as the ASTA.

If the agent has recently graduated from a travel school, he or she may get formal or informal placement assistance through that school. Some schools, for example, announce their graduating classes in industry publications and arrange interviews with potential employers. Others maintain a listing of job openings for their students' use.

Heather Dolstra offers the following advice to aspiring travel agents who are looking for their first jobs in the field: "This is a business where you cannot afford to make many mistakes. Being able to demonstrate to a potential employer that you have skills related to learning, research, and communication will make you a more valuable trainee. It is important in today's marketplace to be able to navigate the Web and do creative searches for information. We have to stay one step ahead of our clients. It doesn't hurt to have personal travel experience to bring to the table, but that experience is wasted if the skill set to pull together information from multiple sources is lacking."

WHERE CAN I GO FROM HERE?

Advancement opportunities for travel agents are somewhat limited. Experienced and skilled agents may advance to the position of *travel office* or *agency manager*. These managers are usually responsible for overseeing other travel agents, generating various reports, keeping track of finances, and generally managing all the activities of the travel agency.

The travel agent with capital, business skills, and a good following of clients might eventually open his or her own travel agency. In order to do this, he or she must generally have approval from supplier organizations, such as the Airlines Reporting Corporation.

In addition to jobs in the regular travel business, a number of travel jobs are available with corporations, automobile clubs, and transportation companies.

Some state and local governments also hire travel professionals for their departments of tourism. Agents with several years of experience may be eligible for a job with one of these organizations.

WHAT ARE THE SALARY RANGES?

Travel agents typically earn a straight salary. Although less common, some agents are paid a salary plus commission or entirely on a commission basis. Salaries of travel agents ranged from less than $18,770 to $47,860 or more in 2008, with an average of $30,570, according to the U.S. Department of Labor. Managers with 10 years of experience may earn more than $65,000 annually. In addition to experience level, the location of the firm is also a factor in how much travel agents earn. Agents working in larger metropolitan areas tend to earn more than their counterparts in smaller cities. The ASTA offers a salary research tool at its Web site; users can search for salary information based on their experience level, state, agency size, and other criteria.

One of the benefits of working as a travel agent is the chance to travel at a discounted price. Major airlines offer special agent fares, which are often only 25 percent of regular cost. Hotels, car rental companies, cruise lines, and tour operators also offer reduced rates for travel agents. Agents also get the opportunity to take free or low-cost group tours sponsored by transportation carriers, tour operators, and cruise lines. These trips, called "fam" trips, are designed to familiarize agents with locations and

Did You Know?

- There were 18,077 Airlines Reporting Corporation-authorized retail travel agencies in the United States as of April 2008.

- Nearly 95 percent of travel agents have used the Internet to conduct research. In 2007, travel agency managers and owners spent an average of 32.5 hours a week using the Internet. The most popular travel segments booked by travel agents online were (in descending order): air travel, tour packages, and hotel reservations.

- Travel agents sell 85 percent of cruises, 70 percent of all tours and packages, 50 percent of all airline tickets, 30 percent of all hotels, and 25 percent of all car rentals.

Sources: *American Society of Travel Agents,* 2008 PhoCusWright Travel Agency Distribution Landscape Report

accommodations so that they can better market them to their clients. Kristle Corcuera was recently invited on a "fam" trip to stay two nights at the Encore Las Vegas. "It was a great experience visiting all of the restaurants, suites, casinos, and the spa," she recalls. "The 'fam' trips are designed to give you a full tour and to get you acquainted with the facility so that you can go home and recommend the location to your clients. They gave me the royal treatment. I was only required to attend a couple meetings in the morning and the rest of the day and night were free to explore and enjoy my stay. The best part was that I was having the time of my life while working—and all for free!"

In addition to travel benefits, most agents also receive a standard benefits package that includes medical insurance and paid holidays, sick days, and vacations. The quality of the benefits package may depend upon the size of the agency, however; some smaller agencies provide less-than-average benefits for their employees. Self-employed travel agents must provide their own benefits.

WHAT IS THE JOB OUTLOOK?

The U.S. Department of Labor predicts that travel agents should have "fair to good job opportunities" through 2016, with those "who specialize in a travel destination, type of traveler, or mode of transportation" expected to have the best opportunities. There are some factors, however, that may negatively influence the growth of jobs for travel agents. Many airlines and other travel suppliers now offer consumers the option of making their own travel arrangements through online reservation services, readily accessible through the Internet. There also are many Web sites that help travelers research, plan, and book trips. With these options, travelers have become less dependent upon agents to make arrangements for them. The travel industry is sensitive to economic changes and political crises that may cause international travel plans

to be postponed. Therefore, the number of job opportunities for agents may fluctuate, depending upon the general political and economic climate.

Despite these negative factors, there will still be demand for travel agents. The public will still rely on them to help plan complex trips and suggest new or offbeat excursions or destinations. Travel agents with advanced educations and who specialize in a particular location, traveler demographic, or other travel area will have the best employment opportunities. "Agencies are always looking for good salespeople," says Joanne Gardner. "Travel experience counts as well, so use what you have. The industry itself has millions of workers if you count hospitality, airlines, suppliers, guides, agents, etc., so there is room for growth in many areas. You need to decide what area you want to enter—hotel management, airlines agency work, etc.—and concentrate on that."

"The employment outlook for travel and hospitality is taking a momentary hit," says Heather Dolstra, "but as the runner up to health care in total number of jobs in this country, prospects for the future are bright. The Internet has automated many travel functions, but there is still a huge need for interpreters and intermediaries. We have found that our overall volume has been only slightly impacted by the Internet after a decade of the public becoming used to this new medium. The bloom is off the rose for many travelers. Everyone has demands on their time. Why try to be a travel agent on top of one's own job?"

Travel Writers

SUMMARY

Definition
Travel writers express, edit, promote, and interpret ideas and facts about travel and the hospitality and tourism industries in written form for newspapers, magazines, books, Web sites (including blogs), and radio and television broadcasts.

Alternative Job Titles
Travel authors
Travel columnists

Salary Range
$28,020 to $53,070 to $106,630+

Educational Requirements
Bachelor's degree

Certification or Licensing
None available

Employment Outlook
About as fast as the average

High School Subjects
English (writing/literature)
Foreign language
Journalism

Personal Interests
Current events
Helping people: personal service
Reading/books
Travel
Writing

Travel writer Margaret Deefholts says that some of the most interesting and rewarding experiences of her career involve meeting and talking with people from other cultures. "It is wonderful to be part of their lives when visiting their lands and their homes," she says. "I have also been privileged to travel on some of the world's most interesting trains, fly by float plane to little-known spots in hidden inlets and coves along the British Columbia coast, and cruise along the fjords of Norway to rarely visited hamlets within and just below the Arctic Circle. Travel writing is hard work for little monetary return, but the payoff is the experience of journeying the world, meeting amazing people, sampling a variety of cuisines, and then seeing your byline and photographs featured in a newspaper or magazine."

WHAT DOES A TRAVEL WRITER DO?

Travel writers report on developments in the travel and hospitality industries; their reports appear in a variety of media—print, broadcast, or online. They may also prepare marketing material for travel industry associations, chambers of commerce, and tourist bureaus. The nature of their work is as varied as the venues for which they write: newspapers, magazines, books, and Web sites and blogs. Some travel writers also appear on television and radio talk shows and documentaries.

Travel staff writers are employed by magazines and newspapers to write feature articles, product reviews, news stories, and columns. First they come up with an idea for an article from their own interests or are assigned a topic by an editor. The topic is of relevance to the particular publication; for example, a writer for a national magazine such as *National Geographic Traveler* might be tasked with writing an article about a felucca trip down the Nile in Egypt; a new four-star hotel that has opened up in Phuket, Thailand; ecotourism opportunities in Patagonia; or a bicycling trip in Croatia. A writer for a regional travel magazine in the Western United States may be assigned an article on bed and breakfasts in the Rockies or great one-day hikes in Yellowstone National Park. A writer for a weekly travel section in a newspaper may be assigned to interview travel experts about money-saving tips on travel or write an article about affordable family-oriented trips that are within a three-hour drive of a given area. A writer for a travel-book publisher might be assigned the task of writing an entire book about a geographic area (Europe), a country (South Africa), a city or state (Minneapolis or Minnesota), a travel trend (ecotourism), or another topic.

After writers receive their assignments, they begin gathering as much information as possible about the subject through library research, interviews, the Internet, observation, and other methods. Most importantly, they visit the place they are writing about to gather firsthand information about the people, culture, and other features of the area, depending on the topic of the article. They keep extensive notes from which they will draw material for their project. They may also take video or photographs, which may be used to illustrate their story.

Once the material has been organized and arranged in logical sequence, writers prepare a written outline. The process of developing a piece of writing is exciting, but it can also involve detailed and solitary work. After researching an idea, a writer might discover that a different perspective or related topic would be more effective, entertaining, or marketable. For example, a writer assigned to write an article about the lighthouses of the Midwest might determine during his or her research and experience in the field that an article on hidden beaches of the Great Lakes might be more interesting to readers.

When working on assignment, writers usually submit their outlines to an editor or other company representative for approval. Then they write a first draft, trying to put the material into words that will have the desired effect on their audience. They often rewrite or polish sections of the material as they proceed, always searching for just the right way of imparting information or expressing an idea or opinion. A manuscript may be reviewed, corrected, and revised numerous times before a final copy is submitted. Even after that, an editor may request additional changes.

Travel columnists analyze news and social issues as they relate to the travel industry. They write about events from the standpoint of their own experience or opinion.

Some travel writers primarily review hotels, restaurants, travel-related products, and travel locales for print publications and television and radio stations. For example, they tell readers and listeners why or why not, in their opinion, they should spend their money to travel to Italy's Cinque Terre (a string of five remote villages on the Italian Riviera); stay in a particular hotel in Vancouver, Canada; or try out a certain new restaurant in the Temple Bar District of Dublin, Ireland.

Writers can be employed either as in-house staff or as freelancers. Pay varies according to experience and the position, but freelancers must provide their own office space and equipment such as computers and fax machines. Freelancers are also responsible for keeping tax records, sending out invoices, negotiating contracts, and providing their own health insurance.

Working conditions vary for travel writers. Although travel writers who work as salaried employees may only work 35 to 40 hours a week, many writers work overtime. A publication that is issued frequently has more deadlines

---

To Be a Successful Travel Writer, You Should...

- enjoy travel
- be highly organized
- enjoy experiencing new cultures and traditions
- speak at least one other foreign language
- have excellent writing skills
- be willing to constantly market yourself to find work (freelance writers)

---

closer together, creating greater pressures to meet them. The work is especially hectic on newspapers, which operate seven days a week. Writers often work nights and weekends to meet deadlines or to cover a late-developing story. Self-employed travel writers work a wide variety of hours based on their assignments.

Most writers work independently, but they often must cooperate with editors, artists, photographers, videographers, and rewriters who may have widely differing ideas of how the materials should be prepared and presented.

Physical surroundings range from comfortable private offices to noisy, crowded newsrooms filled with other workers typing and talking on the telephone. Of course, travel writers also spend a large amount of time traveling to and from a wide variety of destinations, such as bustling cities with millions of people, tiny mountaintop villages, and a wide variety of natural areas.

The work is arduous, but most travel writers are seldom bored. The most difficult element is the continual pressure of deadlines. People who are the most content as travel writers enjoy and work well with deadline pressure and have a deep love of travel and participating in new experiences.

WHAT IS IT LIKE TO BE A TRAVEL WRITER?

Margaret Deefholts is a freelance travel writer based in Canada. Her work has been published in the *Globe and Mail*, the *Vancouver Sun*, The *Georgia Straight,* and a variety of other publications. (Visit http://www.margaretdeefholts-journeys.com to read some of her work.) "I have always been a gypsy at heart and can't remember when I didn't enjoy writing articles, short stories, and poetry," she says. "The two seemed to be a perfect fit for the craft of travel writing. So when I took early retirement from my corporate job, I had the freedom to journey across the globe at short notice, and I invested in a travel writing/marketing/editing course to focus on journalism and photography as it relates to this specialized field.

"Highlights of my career," she continues, "were to have an article published in an anthology of international travel stories (*Literary Trips*) and the publi-

cation of my book, *Haunting India*. It was also very gratifying to have been selected as the winner of the International Travel Writer Award by Tourism Malaysia, which flew me by business class to Kuala Lumpur to accept a trophy at a star-studded tourism gala, and then hosted me on a 10-day trip through the country with all expenses paid."

Margaret says that travel writers need to conduct a fair amount of research before leaving for a trip. "I haunt the library for reference material," she says, "and also scope out the Internet for potential sights that might be of interest to my readers. If there are unusual aspects to a place in the vicinity of my travels—whether it be natural phenomena or a unique museum or a place with little-known historical or cultural significance—I try to fit it into my itinerary. I email those who might be able to offer me an interview, as this always lends a contemporary feel to an article—not to mention affording some intriguing insights. On returning home, I spend at least a couple of days sorting through photographs, diary notes, reference material, video clips, audio clips (interviews), and press kits (if any) before starting on a first draft."

Margaret says that the best part of her job is the "opportunity to travel to parts of the world that I would never have been able to afford to experience without qualifying for sponsorship by various travel industry suppliers (public relations companies handling media and press trips to promote tourism boards, airlines, and hotels). There's a cautionary side to this, however, in that I make it clear to the sponsoring party that this doesn't mean I'll write a cloyingly advertorial article—my first obligation is to my readers, who must be made aware of any negative aspects to a trip. Fortunately that doesn't happen too often."

Tim Leffel is a freelance travel writer based in Nashville, Tennessee. He has worked in the field for 17 years. Tim is the author of *The World's Cheapest Destinations*; *Make Your Travel Dollars Worth a Fortune: The Contrarian Traveler's Guide to Getting More for Less*; and *Traveler's Tool Kit: Mexico and Central America*. His work has been published in numerous travel publications, and he is the editor of the narrative Web publication *Perceptive Travel* (http://www.perceptivetravel.com). (Visit http://www.timleffel.com to learn more about his career and to read some of his work.) Before becoming a travel writer, Tim worked for a music company in marketing and did a lot of writing in that position. "When I left the corporate world and started backpacking around the world," he recalls, "it seemed like an easy transition to become a travel writer. After I got some things published, I kept improving and realized I was pretty good at it, good enough to beat the odds and make more than spending money anyway. I found I was more passionate about this than anything else I could imagine doing full time, so I found a way to make travel writing my regular job.

"Working in this field is a lot more interesting and fun than most other careers in the sense that it involves travel and I am very independent," Tim continues. "I get paid to visit and write about interesting places and talk to interesting people. Like any freelancing career, though, it's a roller coaster financially and I always have to be hustling: the marketing and salesmanship part is in some ways more important than what I actually write in terms of keeping the income stream flowing. A travel writer has to work long hours while on the move because half the job is research and the other half is actually getting the story or book together. Meanwhile, I still have to answer emails and keep the blog posts going, so it's not at all relaxing to be on assignment somewhere; it's more work than when I'm at home in my office."

Tim says that he has visited more countries than he can count, and that he has had too many rewarding and interesting experiences to choose a favorite. "Watching my book *The World's Cheapest Destinations* become a success and seeing it translated into Italian was certainly a thrill, and I'm always gratified when I win some kind of writing award for a book or a story. The most rewarding thing for me, though, has been to take an angle I'm curious about and turn an interesting travel experience into a good, unique article. Whether it's hiking through the mountains of Peru, interviewing Sadhus in India, or taking my daughter along to write about regional amusement parks in the United States, I get jazzed up about shedding a new light on a place and putting out an article that's all my own, something nobody has ever read anywhere else. There are no real barriers to entry in this field, so to rise above the pack I constantly feel like I need to please my toughest critic—myself."

DO I HAVE WHAT IT TAKES TO BE A TRAVEL WRITER?

"Travel writers need to have the ability to meet deadlines," says Tim Leffel. "You have to have a very strong work ethic; I've never met a lazy travel writer, despite what people may picture—an image of a napping guy in a hammock with a cocktail and a notebook. You need to be extra-observant; have a good vocabulary that comes from reading a lot; be naturally curious; be well-versed in geography, art, and history; and be someone who doesn't require a lot of hand-holding to get around and do research. Last, but probably most important, you have to be comfortable selling yourself and your ideas as most travel writers are freelancers, not on staff positions somewhere."

Other important skills for travel writers include creativity, strong communication and research skills, computer literacy, a love of travel, and a desire to learn about new cultures. Travel writers must be confident about their opinions and able to accept criticism from others who may not agree with their views.

One often-overlooked asset for freelance travel writers is skill in marketing. It takes a lot of effort to get published, and travel writers must constantly contact editors with article ideas. "I'm a

terrible marketer," says Margaret Deefholts, "and don't enjoy sending out query letters by the dozen. However, there is no other way to make it into print. One has to learn to accept rejection, or accept the fact that busy editors often don't even bother to respond at all. There are a lot of very gifted and professionally adept travel writers out there, so meticulous writing (spelling, grammar, and scrupulous fact-checking), as well as persistence and timing, is the name of the game. Today the competition for print exposure is even more difficult as the current recession is putting the squeeze on advertising budgets, so newspapers are either shriveling and dying, or putting more material onto their Web sites. In fact, the Internet is the fastest-growing travel writing market available to writers today."

HOW DO I BECOME A TRAVEL WRITER?
Education
High School
While in high school, build a broad educational foundation by taking courses in English, literature, foreign languages, history, general science, social studies, computer science, and typing. The ability to type is almost a requisite for many positions in the journalism field, as is familiarity with computers. If you are interested in becoming a travel writer, you should watch as many television shows or films about travel as possible, as well as read publications and Web sites on travel.

Good Advice

Margaret Deefholts offers the following advice to young people who are interested in becoming travel writers and writers in general:

- Enroll in a travel-writing/marketing and photography course through an accredited college or university if you are serious about pursuing this as a goal.

- Polish and re-polish your material. Print your article out…then read it aloud. Does it have rhythm and flow? Keep descriptions to a minimum (they slow the pace), but don't hesitate to use vivid similes that evoke the senses and take the reader along with you on your journey. Does your story have a compelling hook in the first paragraph? That's all an editor has time to read, so you need to grab his or her attention right away. Keep the article peppy and fast paced.

- Make sure you have photos that enhance the text, and if possible add in a PowerPoint slide show or a video clip (very popular on many travel Web sites).

- Check out the weekly travel articles in several online magazines and publications—the *New York Times,* for example—to get an idea of style, pace, and "hot" spots for travel-hungry tourists. Tailor your query letters accordingly.

- Be persistent in your quest for marketing and publishing your stories. You will eventually succeed, but be patient.

- Freelance travel writers, unfortunately, rarely earn a living wage on a consistent basis. So unless you are on the editorial banner of a large newspaper or magazine, you will need to have other sources of income. Look at diversifying your portfolio by taking on corporate writing (company newsletters, narratives for annual general reports, advertorial copy, etc.), obtaining a regular column in a newspaper (offering advice on a variety of subjects, or political or financial insights), auditioning for a stint on your community TV network (travel program), and taking on teaching/lecturing assignments related to journalism.

Postsecondary Training

Competition for journalistic writing jobs almost always demands the background of a college education. Many employers prefer that you have a broad liberal arts background or majors in English, literature, history, philosophy, or one of the social sciences. Other employers desire communications or journalism training in college. Occasionally a master's degree in a specialized writing field may be required. A number of schools offer courses in journalism, and some of them offer courses or majors in newspaper and magazine writing, publication management, book publishing, and writing for the Internet. If you are interested in travel writing, you might want to consider a major, or at least a minor, in a travel-related area.

Some travel-writing associations offer classes and workshops in travel writing. Taking such classes will be an excellent way to hone your skills and determine if this career is a good fit for your interests and abilities.

In addition to formal course work, most employers look for practical writing experience. If you have worked on high school or college newspapers, yearbooks, or literary magazines, you will make a better candidate. Work for small community newspapers or radio stations, even in an unpaid position, will also help.

Certification or Licensing

No certification or licensing is available for this profession.

Internships and Volunteerships

Many magazines, newspapers, and radio and television stations have summer internship programs that provide valuable training if you want to learn about the publishing and broadcasting businesses. Interns do many simple tasks, such as running errands and answering phones, but some may be asked to perform research, conduct interviews, or even write some minor pieces. You could also volunteer at a local newspaper or at a book publisher. Many travel Web sites allow people to post their travel stories or reviews about travel-related products. Although these don't typically offer pay, writing for one of these sites will give you great experience. You could also start your own blog about one of your favorite travel destinations or other topics.

Hot International Travel Destinations, 2009

Travel agents rate the following international destinations as the most popular for tourists:

1. London, United Kingdom
2. Rome, Italy
3. Paris, France
4. Cancun, Mexico
5. Punta Cana, Dominican Republic
6. Puerto Vallarta, Mexico
7. Barcelona, Spain
8. Venice, Italy
9. Florence, Italy
10. Amsterdam, The Netherlands

Source: American Society of Travel Agents

WHO WILL HIRE ME?

"I landed my first job in the field when an editor who was an instructor in my travel-writing course invited me to send him a query on an article I'd done on India," says Margaret Deefholts. "Over the years we built up a mutually productive relationship, and as my byline began to find recognition, I then branched out with articles sent to local and country-wide daily and weekly newspapers."

Only a small percentage of the approximately 135,000 writers and authors in the United States specialize in writing about travel. About half of salaried writers and editors work for newspapers, magazines, and book publishers; radio and television broadcasting companies; and Internet publishing and broadcasting companies. Outside the field of journalism, writers are also employed by advertising agencies, public-relations firms, and for journals and newsletters published by business and nonprofit organizations, such as professional associations, labor unions, and religious organizations. Other non-journalism employers are government agencies. Other writers work as novelists, short story writers, poets, playwrights, and screenwriters.

The major newspaper, magazine, and book publishers account for the concentration of journalistic writers in large cities such as New York, Chicago, Los Angeles, Boston, Philadelphia, San Francisco, and Washington, D.C. Opportunities with small publishers can be found throughout the country.

It takes many years of experience to gain a high-level position in the field or a byline in a popular magazine. Nearly all writers start out in entry-level positions such as editorial assistant or junior writer. These jobs may be listed with college career services offices, or they may be obtained by applying directly to the employment departments of the individual publishers or broadcasting companies. Graduates who previously served internships with these companies often have the advantage of knowing someone who can give them a personal recommendation. Want ads in newspapers and trade journals are another source for jobs. Because of the competition for positions, few vacancies are listed with public or private employment agencies.

Employers in the field of journalism are usually interested in samples of published writing. These are often assembled in an organized portfolio or scrapbook. Bylined or signed articles are more credible (and, as a result, more useful) than stories whose source is not identified.

Beginning positions as a junior writer usually involve library research, preparation of rough drafts for part or all of a report, cataloging, and other related writing tasks. These are generally carried out under the supervision of a senior writer.

WHERE CAN I GO FROM HERE?

Most salaried travel writers find their first jobs as editorial, production, or research assistants. Advancement may be more rapid in small media companies, where beginners learn by doing a little bit of everything and may be given writing tasks immediately. At large publishers or broadcast companies, duties are usually more compartmentalized. Assistants in entry-level positions are assigned such tasks as research and fact-checking, but it generally takes much longer to advance to full-scale writing duties.

Promotion into higher level positions may come with the assignment of more important articles and stories to write, or it may be the result of moving to another company. Mobility among employees in this field is common. A staff travel writer at a small magazine publisher may switch to a similar position at a more prestigious publication.

Freelance or self-employed writers earn advancement in the form of larger fees as they gain exposure and establish their reputations. Some travel writers, such as Tim Cahill or Pico Iyer, write books about travel-related topics. Others, such as Rick Steves, become household names for their expertise in a particular area (Europe, for Rick Steves) and host television series on travel.

WHAT ARE THE SALARY RANGES?

In 2008 salaried writers working for book publishers had earnings that ranged from less than $28,020 to more than $106,630, according to the U.S. Department of Labor, with an average of $53,070. Writers employed by newspapers and periodicals or in radio and television broadcasting had annual mean earnings of $34,850.

In addition to their salaries, many travel writers earn some income from freelance work. Part-time freelancers may earn from $5,000 to $15,000 a year. Freelance earnings vary widely. Full-time, established freelance writers may earn $75,000 or more a year.

Writers who work as salaried employees receive benefits such as vacation days, sick leave, health and life insurance, and a savings and pension program. Self-employed writers must provide their own benefits.

WHAT IS THE JOB OUTLOOK?

The employment of all writers is expected to increase about as fast as the average rate for all occupations through 2016, according to the U.S. Department of Labor. The demand for writers by newspapers, periodicals, and book publishers is expected to increase. The growth of online publishing on company Web sites and other online services will also demand many talented writers; as a result, those with computer skills will be at an advantage.

People entering the field of writing, especially travel writing, should realize that competition for jobs is extremely keen. Beginners may have marked difficulty finding employment. Of the thousands who graduate each year with degrees in English, journalism, communications, and the liberal arts, intending to establish a career as a writer, many turn to other occupations when they find that applicants far outnumber the job openings available.

"The employment outlook is pretty dim right now for those who want to follow the traditional media path," says Tim Leffel. "Newspapers are on life support, magazines are struggling, and book sales are down. Pay is stagnant or falling across the board. The one bright spot is the Internet, but pay levels there are far below what they have been in print because the landscape is so fragmented for advertising. There may be two dozen travel magazines on the newsstand, but there are probably 2 million travel sites of some kind on the Web. On the other hand, it's much easier now to bypass all that and start your own site, writing about whatever you can specialize in. If you find a good niche and build a following, that probably won't make you rich, but you can at least make a supplemental income without pitching anything to editors. Those who are really good at it and are entrepreneurial can turn that into a real job without sending out any queries."

SECTION 3

Do It Yourself

Katie knew she wanted to work in a hotel. She loved the atmosphere, the energy, and, of course, the glamour, of hotels. (At least that's the way hotels are portrayed in made-for-TV movies.) At 15, Katie didn't have much exposure to different lodging establishments, other than the Holiday Inns her family stayed at on vacations. "Well, there is time enough for serious career moves," Katie thought. "After all, I'm only in high school."

Samantha knew she wanted to work in a hotel. She, like Katie, loved the atmosphere, the energy, and, of course, the glamour, of hotels. Samantha, also 15 years old, didn't have much exposure to different lodging establishments, but was determined to learn as much as possible. "High school is the best time to learn and make good career moves," Samantha thought. "It's never too early to get a head start."

Who do you think is going places? Put your money on Samantha—never before have there been so many possibilities and opportunities available, and in some cases, tailored to high school students interested in entering the hospitality industry. Set yourself apart from the typical Katie. Build your resume—substance not only exists in part-time hotel jobs, but also in setting up your own business, feeding an interest with summer classes, gathering a group of teens with a similar interest in hospitality careers, or earning class credits and pocket money while learning more about the hospitality industry. Education is important, but so are experience and determination. You'll need all three to succeed in this industry.

❑ START A CLUB

If you're lucky enough to attend a school with a hospitality and travel club, all you have to do is join. But if, like most schools, yours does not have such a club, the thing to do is start one. You're probably familiar with a French club or science club, where the goals are to learn more about the subject, to interact with professionals who have experience in the subject, and to have a good time with others who share a common interest. Those are the exact goals of your hospitality and travel club. To learn more, you and your classmates in the club can share books and magazine articles or surf the Internet for information on hospitality and travel careers. To interact with professionals, you can invite a travel writer to speak to your club or ask a concierge to arrange a tour of his or her hotel. And as you work on these and other projects together, you're bound to have a good time!

Every high school has a different procedure for establishing a club, so check with your principal or guidance counselor to see if there are any formal requirements and to learn how it's been done in the past. In every school, however, the most important elements in a new club are members and a faculty adviser. Ask a teacher to serve as your adviser or sponsor, bearing in mind that if school regulations require him or her to attend club meetings, you'll have to plan your schedule around the teacher's. Be flexible when dealing with faculty, and remember that they're even more pressed for time than you are.

You'll also need at least two or three other students who are interested in careers in travel and tourism to get your club started. Ask around school to find out who's interested. If there doesn't seem to be much interest at first, you'll need to create some. Tell others about the club between classes, at lunch, during homeroom. Do your homework and share your knowledge of the opportunities the industry has to offer. Write about it in the school newspaper, hang up posters to announce meetings, and make sure that everyone knows about your new club.

❏ GET A JOB

Having a part-time job, no matter how small or tedious, is a good way to set yourself apart from the rest of the pack. You may really want to work the front desk, but if the only opening is for the switchboard, take the job anyway! Prove yourself a hard worker, and you'll be sitting behind the front counter soon enough. Dan, a junior from Miami, Florida, works part time at a local hotel. He wants a future in hotel management, but for now he tallies and stocks the minibars. What do pop cans and cashew snack packs have to do with running a hotel? Well, looking at the big picture, both jobs require careful planning, record keeping, and judgment. More importantly, working hard at such a routine, entry-level job clearly demonstrates just how committed Dan is to a job in hospitality. Since Dan's supervisors are aware of his hard work and potential, his days of counting candy bars are numbered.

The classified ads in your city's newspaper and on the Internet (at Web sites for hotels, travel agencies, bed and breakfasts, etc.) should list job openings in your area, but you don't have to wait for an opening to be advertised. Target the places where you would like to work. Make a quick phone call to the main number of the hotel or agency to get the name and title of the human resources director. Now sit down and write that person a brief letter. Tell him or her about your interest in a future career in travel and tourism and why working for their business will help you. Politely indicate that you are interested in any part-time, entry-level positions that may be available now or in the next few months. Add that you will call back in a week to follow up. Tell them that you look forward to speaking with them soon. Before you mail your letter, ask your English teacher to proofread it. Include a one-page resume if you have one. After you've mailed your letter, wait a week and make that follow-up call. Introduce yourself, mention your letter, and politely ask if the person has a few minutes to discuss employment opportunities. Even if there is nothing available, if you have made a good impression, you now have a contact in the travel and hospitality business! Ask if it's okay to check back in a few months. Another alternative is to ask your new contact for an information interview to learn more about the industry and its employment opportunities.

You must be prepared to accept the fact that the hotels or travel agencies you contact may have a firm policy about not hiring high school students. Don't let this discourage you from looking elsewhere. The more active you are in looking for a job, the sooner you're likely to find one.

❑ CLEAN UP YOUR ACT

Cleaning chores may be a low priority on many teens' lists of things to do. But if you are one of the few (there are some out there!) who love to clean and are good at it, then you might consider becoming a hotel executive housekeeper. The executive housekeeper holds one of a hotel's highest management positions, overseeing the work of different departments, and taking responsibility for the cleanliness of the entire hotel.

Naturally, the best way to get experience and make valuable contacts is to work part time in the housekeeping department of a hotel. You may not be given actual guest-room assignments; these are usually trusted to trained room attendants. Many large hotels hire high school students during school breaks to help out with the laundry. Don't belittle such a job—a laundry department at a large, urban hotel is responsible for washing everything from bed sheets to staff uniforms. That's about 100,000 pounds of laundry a week! Your local hotel might not have quite that much laundry for you to tackle, but they may have work for you just the same.

Another option is to join a cleaning service as a seasonal or part-time worker. Look in the phone book or the newspaper classifieds for listings of cleaning businesses in your area. Many operations, such as Molly Maid, are franchised businesses, each responsible for hiring and training their crews. You will be taught good cleaning techniques that also save time and effort. Finding ways to streamline cleaning times is an important duty of an executive housekeeper, and this kind of professional experience would be a real asset.

Of course, ambition and initiative are major assets, too, so you might even consider setting up your own cleaning business. Make fliers describing your services—total house cleaning or special cleaning chores—and your fees. Will you charge by the hour or by the chore? If you will bring all cleaning supplies and the vacuum cleaner, then charge a little extra to cover your overhead. If you aren't sure about how much to charge, ask around to get an idea of how much a cleaning person charges for cleaning an average house or apartment. Distribute the flyers to your family, neighbors, and friends. It may be a good idea to advertise in your community or school paper, as well as church bulletins. Professionals never miss an appointment and are always on time, so note your cleaning jobs in an appointment book to keep yourself organized. Don't limit yourself to homes—cleaning opportunities can be found in small businesses, offices, and churches, as well as yards and cars.

❑ TRAVEL ABROAD

If you want to assist travelers—as a tour guide, travel writer, travel agent, or concierge—then one of the best ways to gain professional experience is, of course, by traveling. Yes, vacationing with your family certainly does count as travel. Just by flying off to a Holiday Inn a couple of states away, you should be able to observe and maybe even speak to people working in almost every occupation profiled in this book. But traveling abroad can give you a better, broader experience of the travel and tourism industry—and you don't have to travel with Mom and Dad to get there.

Is your school's French club going to Paris or the Spanish club to Madrid to brush up on their language skills? If so, will you be with them or will you be sitting at home while they meet flight attendants, hotel desk clerks, and other hospitality and travel professionals from around the world? Does the college you plan to attend sponsor a junior year abroad, allowing you to live and study in a foreign country for one or two semesters? Does your church or another local organization sponsor foreign exchanges, where you and a young person from another country swap homes for a month or more? If you answered "yes" to any of these questions—and you probably did—then traveling abroad may be in your not-too-distant future.

❏ GIVE THEM THE GRAND TOUR

Holly Stiel, one of the first female concierges in the United States, started her career selling tickets at a San Francisco tour booth. Before long, tourists were lined up around her booth for reasons other than buying tickets. Tourists approached Holly for directions to city sights and trusted her advice on "must-see" attractions in San Francisco. If you constantly play tour guide for out-of-town relatives, or if your friends ask for suggestions on what to do every Saturday night, then you probably have the makings of a tour guide or hotel concierge.

You might seek out a formal position as a tour guide with your city's tourist office or chamber of commerce. If they don't have such a position, you can volunteer to create one. Research your town's history, bearing in mind the kind of questions people usually ask about buildings, landmarks, and famous people. Plan a tour—anywhere from 30 minutes to two hours, depending on the size of your town—incorporating this history and your knowledge of current happenings. Then arrange to present this tour to officials from the tourist office or chamber of commerce and try to work together to present this tour to the public. Perhaps you could place fliers in the tourist office announcing that you will lead the tours every hour on the hour on Saturday afternoons. This is a lot of work, but one day it could be your career!

If you're looking for something a little less challenging, why not gain experience as a tour guide in a local museum, park, or historical site? The principles of guiding visitors are just the same, and such establishments are usually very welcoming toward young people. You might also be able to act as a tour guide in your own school by showing it off and sharing your inside information with new and visiting students. Speak with your principal about becoming the school's official tour guide; classmates who share your career interests could help you build a club or organization on that premise.

❏ TAKE OFF

No, you won't be flying a commercial jet during the summer between your junior and senior year of high school, but if you are 16 or over and can pass a rigid mandatory physical examination, you may apply for permission to take flying instruction. This is an excellent way to gain experience as a pilot—and impress your friends. This instruction consists of classroom education and flight training from a Federal Aviation Administration (FAA)-certified flight instructor. Before you make your first solo flight, you must get a medical certificate (certifying that you are in good health) and an instructor-endorsed student pilot certificate. In order to get the student pilot certificate, you must pass a test given by the flight instructor. This test will have questions about FAA rules as well as questions about the model and make of the aircraft you will fly. If you pass the test and the instructor feels you are prepared to make a solo flight, the instructor will sign or endorse your pilot certificate and logbook.

You can apply for a private pilot's license when you are 17 years of age and have successfully fulfilled a solo-flying requirement of 20 hours or more, met instrument flying and cross-country flying requirements, and passed a written examination.

Once you are 18 years old and have 250 hours of flying time you can apply for a commercial airplane pilot's license. In applying for this license, you must pass a physical examination and a written test given by the FAA covering safe flight operations, federal aviation regulations, navigation principles, radio operation, and meteorology. You must also submit proof that the minimum flight-time requirements have been completed and, in a practical test, demonstrate flying skill and technical competence to a check pilot. Before you receive an FAA license, you must also receive a rating for the kind of plane you can fly (single-engine, multiengine, or seaplane) and for the specific type of plane.

What's your next stop on the path to working for one of the big airlines, such as United or Southwest? Waiting and practicing your flying. The major airlines require you to be at least 23 years of age to be hired. But that's not too long to wait. In the meantime, you can fly as much as possible, hone your piloting skills, and make valuable contacts with pilots and flight instructors.

❏ FOLLOW THE WRITE PATH

You don't need to wait until you're an adult to begin exploring a career as a travel writer. You can begin writing about your trips right away. You can start a journal or blog; enter writing contests (such as the Norm Stung Youth Writing Competition, which is offered by the Outdoor Writers Association of America); read travel books, magazines, and Web sites; and talk to travel writers about their careers. After reading stories by top travel writers, it may seem daunting to put your thoughts down on paper, but don't give up. You may not have been to Bali, the Great Pyramids, or the Galapagos Islands, but that doesn't mean you can't write a captivating story about a local class trip to Washington, D.C., or a summer sojourn with your parents to Mount Rushmore. There are always new restaurants and attractions opening—and you can help people learn about them through your writing. The key is to use your own voice and descriptive detail to tell readers a story, and provide them with suggestions about what to see, where to stay, what to eat, when to go, and other details typically covered by travel writers. Good writing takes practice, so the sooner you get started, the better.

❑ SHADOW THE PROS

As you already know, the best way to learn if a career is right for you is to experience it, even for a day. Many schools have implemented job-shadowing programs in which students' interests are matched to area businesses. Students from one California high school wanted to learn more about the hotel industry, so they were able to spend a few afternoons "shadowing" actual employees of a nearby hotel. Many teens may want to be in a top position, such as a hotel general manager, but may not be aware of the duties and responsibilities associated with the job. In addition to experiencing a typical day alongside a front desk manager, hotel general manager, and reservation clerk, the students are able to ask questions and help out when possible.

Don't get discouraged if your school lacks a formal job-shadowing program. One option is to work with your principal or counselor to start such a program. It will take a good deal of planning, but, when presented with solid facts, dates, and times, both students and businesses are usually willing to participate. Another option is to strike out on your own. If you want to be a travel agent, ask your relatives and their friends if they can put you in touch with a professional travel agent. You can then ask to be allowed to shadow that travel agent while he or she is at work, perhaps on a Saturday afternoon. Or you can write a letter to a local travel agency explaining your career goals and your desire to shadow an agent. Again, persistence and politeness are key.

❑ CONCLUSION

These are just some ways to explore and train for a career in travel and tourism. The important factor in each of them is a willingness to do something different, something that hasn't been done before in your school or in your community. It won't always be easy. Some of your requests will be rejected and some of your tasks will be nerve-racking, but your efforts will pay off. Trying new things and meeting new people are always to your benefit, though not always in obvious ways. It's worth taking a chance because it's your future.

SECTION 4

What Can I Do Right Now?

Get Involved: A Directory of Camps, Programs, Competitions, and Other Opportunities

Now that you've read about some of the different careers available in travel and tourism, you may be anxious to experience this line of work for yourself, to find out what it's really like. Or perhaps you already feel certain that this is the career path for you and you want to get started on it right away. Whichever is the case, this section is for you! There are plenty of things you can do right now to learn about hospitality and travel careers while gaining valuable experience. Just as important, you'll get to meet new friends and see new places, too.

In the following pages, we list specific programs around the United States that make the travel and tourism industry accessible, in some measure, to high school students. Your opportunities range from flying with an experienced pilot to taking hospitality-related classes at a college. Some will keep you busy during the school year, others will fill up your summer vacations. We've categorized all of the programs in this book for your convenience; it's up to you to decide whether you're interested in one particular type of program or are open to a number of possibilities. The categories into which these programs fall are listed right after the name of the program or organization, so you can skim through to find the listings that interest you most.

❑ THE CATEGORIES

Camps

When you see an activity that is classified as a camp, don't automatically start packing your tent and mosquito repellent. Where academic study is involved, the term *camp* often simply means a residential program that includes both educational and recreational activities. It's sometimes hard to differentiate between such camps and other study programs, but if the sponsoring organization calls it a camp, so do we! Visit the following Web sites for an extended list of camps: http://www.kidscamps.com and http://find.acacamps.org/finding_a_camp.php.

College Courses/Summer Study

These terms are linked because most college courses offered to students your age must take place in the summer, when you are out of school. At the same time, many summer study programs are sponsored by colleges and universities that want to attract future students and give them a head start in higher education. Summer study of almost any type is a good idea because it keeps your mind and your study skills sharp over the long vacation. Summer study at a college offers any number of additional benefits, including giving you the tools to make a well-informed decision about your future academic career. (See the Program Descriptions section of this chapter for more information on these opportunities.)

Competitions

Competitions are fairly self-explanatory, but you should know that there are only a few in this book because travel- and tourism-related competitions on a regional or national level are relatively rare. What this means, however, is that if you are interested in entering a competition, you shouldn't have much trouble finding one yourself. Your school counselor can help you start searching in your area.

Conferences

Conferences for high school students are usually difficult to track down because most are for professionals in the field who gather to share new information and ideas with each other. Don't be discouraged, though. A number of professional organizations with student branches or membership options for those who are simply interested in the field offer conferences. Some student branches even run their own conferences. This is an option worth pursuing because conferences focus on some of the most current information available and also give you the chance to meet professionals who can answer your questions and even offer advice.

Employment and Internship Opportunities

Companies in the travel and tourism industry offer plenty of opportunities for young people looking for their first jobs. Many, if not most, have some kind of entry-level position open to high school students, such as hotel receptionist or desk clerk, theme park worker, or travel agency clerk. (Airlines are, of necessity, a different matter.) Bear in mind that, if you do a good enough job and the group you work for has the funding, this summer's volunteer position could be next summer's job.

Basically, an internship combines the responsibilities of a job (strict schedules, pressing duties, and usually written evaluations by your supervisor) with the uncertainties of a volunteer position (no wages [or only very seldom], no fringe benefits, no guarantee of future employment). That may not sound very enticing, but completing an internship is a great way to prove your maturity, your commitment to a travel- or tourism-related career, and your knowledge and skills to colleges, potential employers, and yourself. Some internships listed here are just formalized volunteer positions; others offer unique responsibilities and opportunities. Choose the kind that works best for you!

Field Experience

This is something of a catchall category for activities that don't exactly fit the other descriptions. But anything called a field experience in this book is always a good opportunity to get out and explore the work of travel and tourism professionals.

Membership

When an organization is in this category, it simply means that you are welcome to pay your dues and become a card-carrying member. Formally joining any organization offers the benefits of meeting others who share your interests and concerns, finding opportunities to take action, and keeping up with current events in the field and in the group. Depending on how active you are, the contacts you make and experiences you gain may help when the time comes to apply to colleges or look for a job.

In some organizations, you may pay a special student rate but receive virtually the same benefits as a regular adult member. Most groups have student branches with special activities and publications. Don't let membership dues discourage you from contacting any of these organizations. Most charge only a nominal fee because they know that students are perpetually short of funds. If the fees are still too much for your budget, contact the group that interests you anyway—they are likely to at least send you some information and place you on their mailing list.

Seminars

Like conferences, seminars are often classes or informative gatherings for those already working in the field, and are generally sponsored by professional organizations. This means that there aren't all that many seminars for young people. But also like conferences, they are often open to affiliated members. Check with various organizations to see what kind of seminars they offer and if there is some way you can attend.

Volunteer Opportunities

Generally speaking, the volunteer opportunities you'll find in this book and around the travel and tourism industry amount to informal internships. As a volunteer, you can expect to work fewer hours and receive less training than an intern—which gives you more flexibility while requiring less commitment from you. This makes volunteering a particularly good option if you just want to explore a variety of potential career paths. Having said that, it is nevertheless true that volunteering can mark the beginning of your career if you want it to.

❏ PROGRAM DESCRIPTIONS

Once you've started to look at the individual listings themselves, you'll find that they contain a lot of information. Naturally, there is a general description of each program, but wherever possible we have also included the following details.

Application Information

Each listing notes how far in advance you'll need to apply for the program or position, but the simple rule is to apply as far in advance as possible. This ensures that you won't miss out on a great opportunity simply because other people got there ahead of you. It also means that you will get a timely decision on your application, so if you are not accepted, you'll still have some time to apply elsewhere. As for the things that make up your application—essays, recommendations, etc.— we've tried to tell you what's involved, but be sure to contact the program about specific requirements before you submit anything.

Background Information

This includes such information as the date the program or organization was established, the name of the organization that is sponsoring it financially, and the faculty and staff who will be there for you. This can help you—and your family—gauge the quality and reliability of the program.

Classes and Activities

Classes and activities change from year to year, depending on popularity, availability of instructors, and many other factors. Nevertheless, colleges and universities quite consistently offer the same or similar classes, even in their summer sessions. Courses like "Introduction to Travel and Tourism" and "Hospitality 101," for example, are simply indispensable. So you can look through the listings and see which programs offer foundation courses like these and which offer courses on more varied topics. As for activities, we note when you have access to recreational facilities on campus, and it's usually a given that special social and cultural activities will be arranged for most programs.

Contact Information

Wherever possible, we have provided the title of the person whom you should contact instead of the name because people change jobs so frequently. If no title is given and you are telephoning an organization, simply tell the person who answers the phone the name of the program that interests you and he or she will forward your call. If you are writing, include the line "Attention: Summer Study Program" (or whatever is appropriate after "Atten-tion") somewhere on the envelope. This will help to ensure that your letter goes to the person in charge of that program.

Credit

Where academic programs are concerned, we sometimes note that high school or college credit is available to those who have completed them. This means that the program can count toward your high school diploma or a future college degree just like a regular course. Obviously, this can be very useful, but it's important to note that rules about accepting such credit vary from school to school. Before you commit to a program offering high school credit, check with your counselor to see if it is acceptable to your school. As for programs offering college credit, check with your chosen college (if you have one) to see if this type of credit is acceptable.

Eligibility and Qualifications

The main eligibility requirement to be concerned about is age or grade in school. A term frequently used in relation to grade level is "rising," as in "rising senior": someone who will be a senior when the next school year begins. This is especially important where summer programs are concerned. A number of university-based programs make admissions decisions partly in consideration of GPA, class rank, and standardized test scores. This is mentioned in the listings, but you must contact the program for specific numbers. If you are worried that your GPA or your ACT or SAT scores, for example, aren't good enough, don't let this stop you from applying to programs that consider such things in the admissions process. Often, a fine essay or even an example of your dedication and eagerness can compensate for statistical weaknesses.

Facilities

We tell you where you'll be living, studying, eating, and having fun during these programs, but there isn't enough room to go into all the details. Some of those details can be important: what is and isn't accessible for people with disabilities, whether the site of a summer program has air-conditioning, and how modern the facilities and computer equipment are. You can expect most program brochures and application materials to address these concerns, but if you still have questions about the facilities, just call the program's administration and ask.

Financial Details

While a few of the programs listed here are fully underwritten by collegiate and corporate sponsors, most of them rely on you for at least some of their funding. The 2009 prices and fees are given here, but you should bear in mind that costs rise slightly almost every year. You and your parents must take costs into consideration when choosing a program. We always try to note where financial aid is available, but really, most programs will do their best to ensure that a shortage of funds does not prevent you from taking part.

Residential vs. Commuter Options

Simply put, some programs prefer that participating students live with other participants and staff members, others do not, and still others leave the decision entirely to the students themselves. As a rule, residential programs are suitable for young people who live out of town or even out of state, as well as for local residents. They generally provide a better overview of college life than programs in which you're only on campus for a few hours a day, and they're a way to test how well you cope with living away from home. Commuter programs may be viable only if you live near the program site or if you can stay with relatives who do. Bear in mind that for residential programs especially, the travel between your home and the location of the activity is almost always your responsibility and can significantly increase the cost of participation.

❏ FINALLY . . .

Ultimately, there are three important things to bear in mind concerning all of the programs listed in this volume. The first is that things change. Staff members come and go, funding is added or withdrawn, supply and demand determine which programs continue and which terminate. Dates, times, and costs vary widely because of a number of factors. Because of this, the information we give you, although as current and detailed as possible, is just not enough on which to base your final decision. If you are interested in a program, you simply must contact the organization concerned to get the latest and most complete information available, or visit its Web site. This has the added benefit of putting you in touch with someone who can deal with your individual questions and problems.

Another important point to keep in mind when considering these programs is that the people who run them provided the information printed here. The editors of this book haven't attended the programs and don't endorse them; we simply give you the information with which to begin your own research. And after all, we can't pass judgment because you're the only one who can decide which programs are right for you.

The final thing to bear in mind is that the programs listed here are just the tip of the iceberg. No book can possibly cover all of the opportunities that are available to you—partly because they are so numerous and are constantly coming and going, but partly because some are waiting to be discovered. For instance, you may be very interested in taking a college course but don't see the college that interests you in the listings. Contact its admissions office! Even if the college doesn't have a special program for high school students, it might be able to make some kind of arrangements for you to visit or sit in on a class. Use the ideas behind these listings and take the initiative to turn them into opportunities.

❑ THE PROGRAMS

American Culinary Federation (ACF)
Conferences/Membership

The ACF, a professional organization for chefs and cooks, offers membership for high school students and culinary enthusiasts. Members receive networking opportunities, access to conferences and conventions, publications, the opportunity to participate in culinary competitions, and discounts on products. High school student members must be between the ages of 16 and 18 and enrolled in a vocational program. The culinary enthusiast category is open to anyone with an interest in the culinary arts.

American Culinary Federation
180 Center Place Way
St. Augustine, FL 32095-8859
800-624-9458, ext. 102
acf@acfchefs.net
http://www.acfchefs.org

American Hotel and Lodging Association
Field Experience

The American Hotel and Lodging Association (AHLA) and its members participate in National Groundhog Job Shadow Day (NGJSD) each February. The NGJSD is a "national campaign that gives young people a new perspective on their studies through hands-on learning and a one-day mentoring experience." In 2008, more than 2,000 restaurants and hotels hosted nearly 20,000 participants. As a participant, you might shadow a hotel manager or reservation clerk or actually work as a hotel employee for a day. This program will provide you with an excellent opportunity to gain firsthand knowledge of the hotel industry, work settings, and careers in the field. Contact the AHLA for more information on the program.

American Hotel and Lodging Association
1201 New York Avenue, NW, Suite 600
Washington, DC 20005-3931
202-289-3100
informationcenter@ahla.com
http://www.ahla.com

American Hotel and Lodging Association Educational Institute

Competitions/Field Experience

The Educational Institute offers two programs for high school juniors and seniors who are interested in working in the hospitality industry: the Lodging Management Program and the START (Skills, Tasks, and Results Training) program.

The Lodging Management Program is a two-year, advanced high school curriculum that combines classroom learning with work experience in the hospitality industry. More than 7,000 students at 575 high schools have participated in the program. The program has six units: Overview of Lodging Management, The Front Office, Housekeeping, Leadership and Management, Marketing and Sales, and Food and Beverage Service. Graduating seniors who pass examinations and work in the lodging industry for at least 160 hours receive the certified rooms division specialist designation. Program participants also compete for lodging scholarships at a national competition. Past competition activities have included a room inspection (where competitors had to inspect an actual hotel room and find 10 things that were wrong—no pillowcases, dirty linens, broken television, etc.), a knowledge-bowl competition, and sales/marketing or food/beverage case-study presentations.

The START program is a one-year or 180-classroom-hour curriculum (Hospitality Orientation, Rooms Division, and Food and Beverage Division) that "provides training for hospitality positions in the rooms and food and beverage divisions of a lodging operation, including front desk, reservations, housekeeping, bell services, restaurant service, banquet setup and service, and more. Guest service, professionalism, and career exploration are other important components of this program." It is geared toward at-risk youth, welfare-to-work program participants, and other groups who might need extra help building the skills necessary for success in the workplace. High school students in all grades are eligible, although the foundation suggests that this program is most appropriate for freshmen or sophomores who plan to participate in the Lodging Management Program as upperclassmen. Students who complete the program receive a voucher that waives the fees for a hospitality skills certification in one of the 12 areas covered in the START curriculum. This gives them a good start to building a career in the hospitality industry.

American Hotel and Lodging Association Educational Institute
800 North Magnolia Avenue, Suite 300
Orlando, FL 32803-3261
800-349-0299
http://www.lodgingmanagement.org

Big Apple Greeter

Volunteer Opportunities

Big Apple Greeter is a nonprofit organization whose mission is to enhance New York City's image by helping visitors discover its hidden treasures. If you live in New York and love everything about it, perhaps there's a tour guide hidden in you. Volunteer tour guides welcome visitors to all five boroughs and show them that the big city can have small-town charm. Greeters personally take visitors to neighborhoods, cultural attractions, historic sites, and hot spots that only a New Yorker would know. If you'd prefer a behind-the-scenes role, the opportunities include responding to inquiries from visitors, matching greeters with visitors, and providing general office and data-entry assistance. This is a great way to gain experience in the field of hospitality and tourism. Get all the details by phone, email, or mail, or check out Big Apple Greeter's Web site, where you can apply for volunteer positions online. If you don't live in New York City, perhaps there is a similar service in your city or town. Check with your local government or convention bureau for opportunities in your area.

Big Apple Greeter
One Centre Street
New York, NY 10007-1602
212-669-7308 (volunteer info)
volunteerdepartment@bigapple
    greeter.org
http://www.bigapplegreeter.org

Camp Chi
Camps

Camp Chi, located near the beautiful Wisconsin Dells, features many activities, including art, athletics, cooking, media, outdoor adventure, performing arts, and water sports. Students interested in the media option can work on the camp newspaper. Campers work as writers, reporters, editors, and designers for the newspaper and learn the fundamentals of publishing a newspaper from instructors.

In addition to all the activities, the camp, which is operated by the Jewish Community Centers of Chicago, has a heated swimming pool, a spring-fed lake with waterfront activities, a climbing and rappelling wall, a roller hockey arena, rope courses, six tennis courts, and an animal farm. The staff-to-camper ratio is one to three. Camp Chi is for students ages nine to 16. You stay in cabins with built-in bunk beds. If you're 14 to 16 years old, Camp Chi offers a separate village just for teens. Cost of the camp ranges from $1,235 to $6,710, depending on age level and program. This cost includes everything but transportation to the site. Visit Camp Chi's Web site for more information.

Camp Chi
Summer Office:
PO Box 104
Lake Delton, WI 53940-0104
608-253-1681
http://www.campchi.com

Winter Office:
5050 Church Street
Skokie, IL 60077-1254
847-763-3551
http://www.campchi.com

College and Careers Program at Rochester Institute of Technology
College Courses/Summer Study

The Rochester Institute of Technology (RIT) offers its College and Careers Program for rising high school seniors who want to experience college life and explore career options in engineering sciences; computing; science and mathematics; business; liberal arts; art, design, and crafts; and photography. The program, in existence since 1990, allows you to spend a Friday and Saturday on campus living in the dorms and attending up to four sessions in the career areas of your choice. In each session, participants work with RIT students and faculty to gain hands-on experience in the topic area. One recent session was titled Hospitality & Service Management: Super Resorts of Tomorrow. Participants imagined hotels and resorts of the future and participated in a variety of fun, hands-on activities. Other recent classes that might be of interest include International Business: Impacts and Issues in the New Global Economy; Information Systems: The Power Behind a Successful Business; Marketing: Will Your Idea Sell?; and General Management: Why Are Creativity and Innovation Essential in Business? The program is held twice each summer, usually once in mid-July and again in early August. The registration deadline is one week before the start of the program, but space is limited and students are accepted on a first-come, first-served basis. For further information about the program and specific sessions on offer, contact the RIT admissions office.

College and Careers Program

Rochester Institute of Technology
Office of Admissions
60 Lomb Memorial Drive
Rochester, NY 14623-5604
585-475-6631
http://ambassador.rit.edu/careers/
  sessions.php

Collegiate Scholars Program at Arizona State University

**College Courses/Summer Study/
Employment and Internship
Opportunities**

The Collegiate Scholars Program allows high school students to earn college credit during summer academic sessions. Students get the opportunity to explore careers and interact with college professors, as well as receive access to internships, mentoring programs, and research opportunities. Arizona high school seniors may apply, and they are evaluated for admission based on their "high school GPA and/or class rank, test scores, high school schedules, and involvement in other programs offering college credit." Some of the courses that will be of interest to students who would like to explore the wide array of career options in travel and tourism include Private Pilot Ground School, Air Traffic Control, Aviation Meteorology, Math for Business, Statistics, Accounting, Writing for the Professions, Public Speaking, Cultural Diversity, Introduction to Southeast Asia, and various foreign languages courses. Contact the Collegiate Scholars executive coordinator for information on program costs and other details.

Arizona State University

Collegiate Scholars Program
Attn: Executive Coordinator
480-965-2621
mark.duplissis@asu.edu
http://promise.asu.edu/csp

Early Experience Program at the University of Denver

College Courses/Summer Study

The University of Denver invites academically gifted high school students interested in a variety of subjects to apply for its Early Experience Program, which involves participating in university-level classes during the school year and especially during the summer. Recently offered courses include Exploring the World of Hospitality; Survey of Hospitality; Managing the Restaurant Operation; Managing the Lodging Operation; Principles of Tourism; Selling the Hospitality Experience; Hospitality Information Systems and Technology; and various business, mathematics, writing, and foreign language courses. This is a commuter-only program. Interested students must submit a completed application (with essay), official high school transcript, standardized test results (PACT/ACT/PSAT/SAT), a letter of recommendation from a counselor or teacher, and have a minimum GPA of 3.0. Tuition is approximately $1,850 per four-credit class. Contact the Early Experience Program coordinator for more information.

University of Denver

Center for Innovative and
    Talented Youth
Early Experience Program
Attn: Coordinator
1981 South University Boulevard
Denver, CO 80208-0001
303-871-3408
http://www.du.edu/city/programs/
    year-round-programs/early-
    experience-program

Earthwatch Institute

Conferences/Field Experience/Employment and Internship Opportunities/Volunteer Opportunities/Membership

Earthwatch Institute is an organization for people whose spirit of adventure is as great as their commitment to the earth's well-being. A nonprofit membership organization founded in 1971, Earthwatch's major activity is linking volunteers with scientific research expeditions that need them. There are about 130 different expeditions every year, covering all continents but Antarctica, each lasting anywhere from five days to almost three weeks. If you are 16 or 17, you can join a Teen Team and participate in an expedition researching Costa Rican caterpillars, for example, or Australia's fossil forests. Whichever expedition you choose, you work with five to 10 other people under the guidance of a research scientist (often a university professor working in his or her field of expertise).

Living and working conditions vary widely among the expeditions; you might stay in a hotel or a tent, or remain at one site or hike to several locations while carrying a heavy backpack. Expenses also vary widely, from about $199 to $4,000, depending on travel, accommodation, eating arrangements, and other necessary provisions. Earthwatch reminds potential volunteers, however, that your payment of expenses (along with the donation of your time) is really an investment in environmental research. Of course, you're also investing in your own future. With so many expeditions to choose from, you'll be able to gain experience in career fields ranging from ecology to national park service, and from natural history to wildlife preservation. Contact Earthwatch for its annual catalog listing all the details.

Even if you're not up for one of their demanding expeditions, Earthwatch invites you to become a member at the standard rate of $35 per year. You can also attend Earthwatch's annual conference or apply for an internship at its offices in Oxford, England; Melbourne, Australia; or Tokyo, Japan. Contact the institute for more information.

Earthwatch Institute
Three Clock Tower Place, Suite 100
PO Box 75
Maynard, MA 01754-2549
800-776-0188
info@earthwatch.org
http://www.earthwatch.org

Environmental Studies Summer Youth Institute at Hobart and William Smith Colleges
College Courses/Summer Study

Hobart and William Smith Colleges sponsor the Environmental Studies Summer Youth Institute (ESSYI) for rising high school juniors and seniors. Academically talented students are invited to participate in this examination of environmental issues from scientific, social, and humanistic perspectives; the institute will be of interest to aspiring adventure-travel specialists, tour guides, travel writers, and others interested in employment in the travel and tourism industry. Running for two full weeks in July, the ESSYI comprises classroom courses, laboratory procedures, outdoor explorations, and plenty of time to discuss and think about integrating these many approaches to understanding the environment. Lectures encompass ecology, philosophy, geology, literature, topography, and art, among other areas of study, and are conducted by professors from Hobart and William Smith Colleges. Your study of the environment and how humans relate to it also includes field trips to such places as quaking bogs, organic farms, the Adirondack Mountains, and Native American historical sites. Participants also make use of the *HMS William F. Scandling*, the colleges' 65-foot research vessel, as they explore the ecology of nearby Seneca Lake. ESSYI students live on campus and have access to all of the colleges' recreational facilities. Those who complete this intellectually and physically challenging program are awarded college credit. The fee for the program is $2,200. For information on financial aid and application procedures, contact the institute director.

Environmental Studies Summer Youth Institute
Hobart and William Smith Colleges
Attn: Director
Geneva, NY 14456-3397
315-781-4401
essyi@hws.edu
http://academic.hws.edu/enviro

Experimental Aircraft Association
Camps/Employment and Internship Opportunities/Field Experience/Membership/Volunteer Programs

The Experimental Aircraft Association (EAA) seeks to get young people interested and involved in aviation via a variety of programs and activities.

Its free Young Eagles Flight Program matches young people between the ages of eight and 17 who are fascinated by flight with adult pilots eager to share their enthusiasm for aviation. Young Eagles actually fly with the pilots. Flights last 15 to 20 minutes. More than 1.3 million young people have participated in the program since 1992. Opportunities in this program are available throughout the United States.

Students between the ages of 12 and 18 can take part in the EAA Air Academy, where they spend a week immersed in the world of aviation in Oshkosh, Wisconsin. Participants learn about aviation through classes, hands-on activities, flight simulation, a flight in an actual airplane, and other activities. Participants stay in the Air Academy Lodge, which has 12 bunkrooms with four bunks per room and shared bathroom facilities, recreational facilities (volleyball court, basketball court, slate pool table, Foosball, Ping-Pong, and televisions), a library, and a gazebo with fire pit. Tuition for those ages 12–13 is $600 for EAA members and $675 for nonmembers; ages 14–15, $800 (members) and $875 (nonmembers); ages 16–18, $1,000 (members) and $1,075 (nonmembers). Financial aid is available.

The EAA also offers summer internships for members or those who have been recommended by current members. Applicants must be committed to supporting association activities and youth programs. Responsibilities "may range from assisting in research and filing in the Boeing Aeronautical Library to moving and securing aircraft and minor maintenance of buildings, grounds, and aircraft." Internships are typically available to college-level students. Volunteer opportunities also are available.

Finally, membership is available for students. Those age 17 or under can join in the student category ($10) and receive "access to an exclusive student members-only Web site featuring more online tools and resources, career planning information, and applications for aviation scholarships." They also receive a Student Membership Kit, which contains an X-Plane flight simulator demo; EAA Aviation Highlights DVD, *Reach for the Sky*; and a Whitewings glider.

Experimental Aircraft Association

3000 Poberenzy Road
Oshkosh, WI 54902-8939
800-564-6332
http://www.eaa.org

EAA Young Eagles

PO Box 2683
Oshkosh, WI 54903-2683
920-426-6114
http://www.youngeagles.org

EAA Air Academy

PO Box 2683
Oshkosh, WI 54903-2683
920-426-6114
airacademy@eaa.org
http://www.youngeagles.org/
    programs/airacademy/

Exploration Summer Programs: Senior Program at Yale University
College Courses/Summer Study

Exploration Summer Programs (ESP) has been offering academic summer enrichment programs to students for more than three decades. Rising high school sophomores, juniors, and seniors can participate in ESP's Senior Program at Yale University. Two three-week residential and day sessions are available. Participants can choose from more than 80 courses. Recently offered courses include The Write Stuff—Creative Writing; Go to Press!—Print Journalism; From Deserts to Dubai—Arabic Language & Culture; Can You Hear Me Now?—Advertising + Marketing; Speak Easy!—Public Speaking; and Explo Apprentice—Introduction to Business Management. All courses and seminars are ungraded and not-for-credit. In addition to academics, students participate in extracurricular activities such as tours, sports, concerts, weekend recreational trips, college trips, and discussions of current events and other issues. Basic tuition for the Residential Senior Program is approximately $4,555 for one session and $8,390 for two sessions. Day session tuition ranges from approximately $2,100 for one session to $3,820 for two sessions. A limited number of need-based partial and full scholarships are available. Programs are also available for students in grades four through nine. Contact ESP for more information.

Exploration Summer Programs
932 Washington Street
PO Box 368
Norwood, MA 02062-3412
781-762-7400
http://www.explo.org

Federal Aviation Administration
Employment and Internship Opportunities

The Federal Aviation Administration (FAA) offers the Summer Employment Initiative for students ages 16 and older. Students in the program work as clerks, engineering technicians, and in other positions at FAA headquarters (Washington, D.C.) and field locations throughout the

United States. These jobs provide an introduction to typical work settings and job specialties with the FAA. Applicants must be U.S. citizens and have a GPA of at least 2.0 on a scale of 4.0. The duration of the job will not exceed six months (April through September), and part-time and full-time employment is available. Contact the FAA for more information on the initiative.

Federal Aviation Administration
800 Independence Avenue, SW
Washington, DC 20591-0001
866-835-5322
http://www.faa.gov/about/office_
  org/headquarters_offices/ahr/jobs_
  careers/student_programs/
  summer_employment/summer_
  initiative

High School Honors Program/ Summer Challenge Program/ Summer Preview at Boston University
College Courses/Summer Study

Three summer educational opportunities are available for high school students. Rising high school seniors can participate in the High School Honors Program, which offers six-week, for-credit undergraduate study at the university. Students take two for-credit classes (up to eight credits) alongside regular Boston College students, live in dorms on campus, and participate in extracurricular activities and tours of local attractions. Classes are available in more than 50 subject areas, including hospitality (recent classes include Introduction to the Hospitality Industry and Hospitality Field Experience), accounting, anthropology, creative writing, foreign languages (Arabic, Chinese, French, German, Ancient Greek, Hebrew, Italian, Japanese, Korean, Latin, Spanish), management, marketing, and public relations. The program typically begins in early July. Students who demonstrate financial need may be eligible for financial aid. Tuition for the program is approximately $4,120, not including registration/program/application fees ($550) and room and board options ($1,897 to $2,055).

Rising high school sophomores, juniors, and seniors in the University's Summer Challenge Program learn about college life and take college classes in a noncredit setting. The program is offered in three sessions. Students choose two seminars (which feature lectures, group and individual work, project-based assignments, and field trips) from a total of 15 available programs, including Mass Communication (which covers film, television, advertising, public relations, and journalism), Business: From the Ground Up, Introduction to Chinese, and Creative Writing. Students live in dorms on campus and participate in extracurricular activities and tours of local attractions. The cost of the program is approximately $3,070 (which includes tuition, a room charge, meals, and sponsored activities).

Rising high school freshman and sophomores can participate in one-week Summer Preview Programs. This noncredit, commuter program introduces students to college life and a particular area of study, including the study of writing. Students in the Learning the Art of Writing program will learn about various writing styles and techniques via a workshop environment. The cost of the program is $1,100 (which includes tuition, textbooks, lunch, and activities). No financial aid is available.

Boston University High School Programs

755 Commonwealth Avenue,
  Room 105
Boston, MA 02215-1401
617-353-1378
buhssumr@bu.edu
http://www.bu.edu/summer/
  high-school-programs

The International Ecotourism Society
Conferences/Employment and Internship Opportunities/ Membership/Volunteer Programs

This nonprofit organization is "committed to helping organizations, communities, and individuals promote and practice the principles of ecotourism." It offers a traveler membership category for ecotourists and supporters of ecotourism. Benefits include discounts on select ecolodges and operators, society publications, and access to volunteer, internship, and job opportunities.

Students interested in ecotourism can volunteer at organization events such as conferences and meetings. Volunteers at past events were tasked with "assisting speakers with their presentations, taking minutes, guiding participants, distributing promotional materials, and helping with receptions and other networking functions." Students can also participate in voluntourism activities (which the society defines as "tourism activities involving various types of volunteering"). Visit the society's Web site to learn more about voluntourism opportunities with its members.

The International Ecotourism Society
PO Box 96503, #34145
Washington, DC 20009-6503
202-506-5033
http://www.ecotourism.org

Intern Exchange International
Employment and Internship Opportunities

High school students ages 16 to 18 (including graduating seniors) who are interested in gaining real-life experience in travel and tourism can participate in a month-long summer internship in the United Kingdom. Young people considering careers in the hospitality industry can pursue an internship in hotel management, working alongside the professionals at The Lanesborough, Westbury, Four Seasons, or the Langham hotels in London. You'll learn about all the major hotel concerns, from the front desk to the kitchen to the guest rooms. Internships are also available in business/finance, culinary arts, public relations/marketing/advertising, publishing, and other fields. Additionally, a Print & Broadcast Journalism Media and Design Workshop is available or those between the ages of 15 and 18. The cost of either program is approximately $7,335 plus airfare; this fee includes tuition, housing (students live in residence halls at the University of London), breakfast and dinner daily,

housekeeping service, linens and towels, special dinner events, weekend trips and excursions, group activities including scheduled theatre outings, and a Tube Pass. Contact Intern Exchange International for more information.

Intern Exchange International, Ltd.
2606 Bridgewood Circle
Boca Raton, FL 33434-4118
561-477-2434
info@internexchange.com
http://www.internexchange.com

Internship Connection
Employment and Internship Opportunities

Internship Connection provides summer or "gap year" internships to high school and college students in Boston, New York City, and Washington, D.C. Internships are available in hospitality/hotel management. As part of the program, participants learn how to create a resume, participate in a job interview, and develop communication and personal skills that are key to success in the work world. They also get the chance to make valuable contacts during their internships that may help them land a job once they complete college. The program fee for interns in New York or Washington is $2,500, and $2,000 for those in Boston. Contact Internship Connection for more information.

Internship Connection
17 Countryside Road
Newton, MA 02459-2915
617-796-9283

carole@internshipconnection.com
http://www.internshipconnection.
com

Learning for Life Exploring Program
Field Experience

Learning for Life's Exploring Program is a career exploration program that allows young people to work closely with community organizations to learn life skills and explore careers. Opportunities are available in the following programs: Arts & Humanities, Aviation, Business, Communications, and other fields. Each program has five areas of emphasis: Career Opportunities, Service Learning, Leadership Experience, Life Skills, and Character Education. As a participant in the Aviation program, you might "take orientation flights in military transports, helicopters, gliders, or single-engine general aviation aircraft; visit Air Force bases, aviation museums, air shows, or Federal Aviation Administration facilities; learn to preflight an aircraft; and take pilot training ground-school classes."

To be eligible to participate in this program, you must be between 15 and 21 years of age or be 14 and have completed the eight grade.

To find a Learning for Life office in your area (there are more than 300 throughout the United States), contact the Learning for Life Exploring Program.

Learning for Life Exploring Program
1325 West Walnut Hill Lane
PO Box 152079

Irving, TX 75015-2079
972-580-2433
http://www.learningforlife.org/
exploring

National Academy Foundation
Employment and Internship Opportunities/Field Experience

If your high school is pursuing school-to-work initiatives, you may have already heard of the National Academy Foundation (NAF). A nonprofit foundation, it was established in 1982 to promote partnerships between businesses and public high schools. Under NAF's guidance, these partnerships grow into Academies, or schools within schools, that focus on such career fields as finance, engineering, information technology, and—yes—hospitality and tourism. If you have the opportunity to participate in such an Academy at your school, give it some serious consideration. Participants in NAF's programs have generally enjoyed great success in college and in their careers because they are well prepared for the world of work. Students involved in the Academy of Hospitality & Tourism participate in six- to eight-week paid internships with hospitality- or travel-related activities, job shadowing, field trips, and mentoring. They may also have the opportunity to earn college credits while in high school. If your high school does not have an Academy, it might be worth your while to give your principal information about the National Academy Foundation. Even if he or she is interested, it will take some time to set up a partnership, but it should be worth the wait.

National Academy Foundation
39 Broadway, Suite 1640
New York, NY 10006-3081
212-635-2400
http://www.naf.org

National High School Institute at Northwestern University
College Courses/Summer Study

The National High School Institute is the nation's oldest university-based program for outstanding high school students. It was established in 1931. The month-long program offers the following courses: debate, speech, journalism, music, film and video, and theatre arts. Students in the journalism program learn reporting skills and participate in hands-on writing workshops. They also attend seminars given by journalism professors and take field trips to media sites in the Chicago area. The student-to-teacher ratio for these programs is 6 to 1. Applicants for the journalism section must be rising seniors, rank academically in the top 25 percent of their class, and "meet a high standard of character, dependability, and intelligence." A variety of extracurricular activities are also available to students in the program, including tours, movies, shopping, sing-alongs, and outings to sporting and cultural events. Students live on campus in university residence halls, where they also take their meals. Costs range from approximately $2,850 to $6,500 depending on the program (tuition for the journalism program is about $4,550); these amounts include tuition, room, board, health services, field trips, and group events. Scholarships are available. The early admission deadline is typically in the beginning of March, while the regular admission deadline is in late March or early April. Visit the program's Web site for more information.

Northwestern University
National High School Institute
617 Noyes Street
Evanston, IL 60208-4165
800-662-6474
nhsi@northwestern.edu
http://www.northwestern.edu/nhsi

National Park Service Student Educational Employment Program
Youth Programs/Employment and Internship Opportunities/Field Experience/Volunteer Programs

The federal government's Student Educational Employment Program is available to high school, college, and professional degree students. Participants are paid a salary and gain valuable work experience while attending school, which may lead to future employment with the National Park Service (NPS) or other federal agencies after graduation. Applicants must be U.S. citizens or residents of American Samoa or Swains Islands. For further information visit http://www.opm.gov/employ/students.

The NPS offers more than 25 programs for people between the ages of five and 24. The programs, such as the Youth Conservation Corps and Public Land Corps, will help educate you about the environment while you work with conservation workers to improve national parks. Visit the NPS Web site to learn about the wide range of programs that are available and to view photos of past projects.

You can also protect and preserve America's natural and cultural heritage by becoming a park volunteer. You might work as a volunteer at a visitor center in Acadia National Park or a horse center volunteer at Rock Creek Park, help out in the office at Big Cypress National Preserve, or perform a variety of other tasks. Visit the NPS Web site to search for volunteer opportunities by state and national park.

National Park Service
U.S. Department of the Interior
1849 C Street, NW
Washington, DC 20240-0001
202-208-6843
http://www.nps.gov/gettinginvolved

Outdoor Writers Association of America
Competitions
Students in grades six through 12 can participate in the Norm Stung Youth Writing Competition. Submissions must be about outdoors-oriented themes (nature, ecology, hiking, boating, camping, fishing, hunting, canoeing, etc.) and have already been published in a school-related or commercial newspaper, newsletter, magazine, collection, or other publication. There are contest categories for prose and poetry in two divisions: junior (grades six through eight) and senior (grades nine through 12). Winners receive a cash prize and a certificate of honor. Visit the association's Web site for more information.

Outdoor Writers Association of America
121 Hickory Street, Suite 1
Missoula, MT 59801-1896
406-728-7434
http://owaa.org/youth_writing.htm

Professional Association of Innkeepers International (PAII)
Membership/Seminars
The association offers an aspiring innkeeper membership option, although its cost ($199) suggests that it is geared more toward young professionals or career changers. Membership benefits include discounts on PAII resources as well as B & B-related products and services; access to PAII seminars, workshops, and trade shows; opportunities for networking; and a subscription to *Innkeeping Quarterly*.

Professional Association of Innkeepers International
207 White Horse Pike
Haddon Heights, NJ 08035-1703
800-468-7244
http://www.innkeeping.org/

SkillsUSA
Competitions
SkillsUSA offers "local, state and national competitions in which students demonstrate occupational and leadership skills." Students who participate in its SkillsUSA Championships can compete in categories such as Advertising Design, Aviation Maintenance Technology, Commercial Baking, Culinary Arts, Customer Service, Entrepreneurship, Food and Beverage Service, and Job Interview. SkillsUSA works

directly with high schools and colleges, so ask your guidance counselor or teacher if it is an option for you. Visit the SkillsUSA Web site for more information.

SkillsUSA
14001 SkillsUSA Way
Leesburg, VA 20176-5494
703-777-8810
http://www.skillsusa.org

Summer at Georgetown University for High School Students
College Courses/Summer Study

Academically gifted high school students can earn up to 12 college credits by participating in Georgetown University's Summer College. Rising sophomores, juniors, and seniors may apply. More than 100 courses are available, including Accounting, Fundamentals of Finance, Fundamentals of Business, Introduction to Business, Public Speaking, World History, and foreign languages (Arabic, Chinese, French, German, Italian, Japanese, Persian, Spanish). Tuition is $1,018 per credit hour. Other costs include a pre-college fee ($398 per session), room ($782), and a meal plan ($864). Financial aid is available "for exceptional students who can document financial need." Additionally, Georgetown offers several seminars that will be of interest to readers of this book, including Fundamentals of Business: Leadership in a Global Economy, Journalism Workshop, Leadership & Ethics, and Information Technology. Costs and program lengths vary for each seminar.

Residential and commuter options are available for all programs. Students who live on campus stay in air-conditioned residence halls. Access to laundry facilities is provided. In their off hours, students can attend dances, movie nights, ice cream socials, and other activities, as well as explore the campus and the Washington, D.C., area. Contact the university for more information.

Georgetown University
Summer Programs for High School
  Students
Box 571006
Washington, DC 20057-1006
scsspecialprograms@georgetown.edu
http://scs.georgetown.edu/programs/
  113/summer-programs-for-high-
  school-students-summer-college

Summer College Programs for High School Students at Cornell University
College Courses/Summer Study

Rising high school juniors and seniors and recent graduates can participate in Cornell University's Summer College, which offers three- and six-week classes for college credit. More than 20 courses are available. Students who are interested in travel and tourism can take Hotel Operations Management, a three-week course. You will learn about the hospitality industry by attending classes, participating in seminars, listening to guest speakers, and engaging in other hospitality-related activities. You will study the operating strategies and structures of some of the largest hotels and food and beverage companies to learn what makes them successful; learn about the various hotel departments and jobs available in them, and use the CHESS

Hotel Simulation to operate a virtual 250-room hotel. You will interact closely with professors, hospitality professionals, and fellow students. You must bear in mind that all Summer College classes are regular undergraduate courses condensed into a very short time span, so they are especially challenging and demanding. Program participants live in residence halls on campus and enjoy access to campus facilities. The cost for the program is $5,310 (which includes room and board). Applications are typically due in late April, although Cornell advises that you submit them well in advance of the deadline; those applying for financial aid must submit their applications by April 1. Further information and details about the application procedure are available from the Summer College office.

Cornell University Summer College for High School Students

B20 Day Hall
Ithaca, NY 14853-2801
607-255-6203
http://www.sce.cornell.edu/sc/ programs

Summer Scholars at the University of Richmond

College Courses/Summer Study

The Summer Scholars Program seeks to provide high school students a "realistic, firsthand experience of college, including the challenges and rewards that come with it, all while experiencing 'life on campus.'" Rising juniors and seniors are eligible to participate in this three-week, for-credit residential program. Approxi-

mately five courses are offered each summer, including one called Into the Green: Explorations of Text and Trail. This class takes a look at the art of writing about nature. Participants will study nonfiction and fiction nature writing "to see what others have learned, to discuss various writing styles, and to think more deeply about our own relationships to nature and technology." Applicants must have a competitive grade point average and enjoy intellectual stimulation and academic challenges. Program participants stay in air-conditioned residence halls and have access to study lounges, vending machines, and laundry facilities. The cost of the program is $4,200 (which includes tuition, textbooks and classroom supplies, residence hall lodging, a meal plan, and extracurricular activities). Financial assistance is available. Students can use on-campus facilities such as the library, computer labs, and a sports center. The application deadline is typically in early May. Contact the director of summer programs for more information.

University of Richmond School of Continuing Studies

Director of Summer Programs
Richmond, VA 23173
804-289-8382
dkitchen@richmond.edu
http://summer.richmond.edu/scholars

Summer University at Johns Hopkins University

College Courses/Summer Study

Rising high school juniors and seniors who are interested in getting a jump on college can participate in the Summer University program at Johns Hopkins University. Participants live on Hopkins' Homewood campus for five weeks beginning in early July. Classes leading to college credit are available in more than 30 programs. Readers of this book might want to enroll in Introduction to Business, Financial Accounting, Principles of Marketing, Technical Communication, and various foreign language (Arabic, French, German, and Spanish) courses. Students who live in the greater-Baltimore area have the option of commuting. Applicants must submit an application form, essay, transcript, two recommendations, and a non-refundable application fee (rates vary by date of submission). Tuition for residential students is $6,300 (for two courses, room and board, and up to six credits). Commuter students pay $630 per credit hour (books, supplies, meals, and special activities are not included in this price). Contact the Office of Summer Programs for more information.

Johns Hopkins University
Pre-College Program
Office of Summer Programs
Shaffer Hall, Suite 203
3400 North Charles Street
Baltimore, MD 21218-2685
800-548-0548
summer@jhu.edu
http://www.jhu.edu/~sumprog

Summer Youth Explorations at Michigan Technological University
College Courses/Summer Study
Michigan Technological University (MTU) offers the Summer Youth Explorations program for students in grades

six through 12. Participants attend one of five weeklong sessions, choosing either to commute or to live on campus. Students undertake an Exploration in one of many career fields through laboratory work, field trips, and discussions with MTU faculty and other professionals. Past Explorations include Creative Writing in Nature and History; Digital Photography; Entrepreneurship: Start a Business While in High School; Journalism; Learning to Lead: A Leadership Introduction for Everyone; and Create Your Future! What Do You Want to Be? How Do You Do It? The cost of the Summer Youth Program is $650 for the residential option, $395 for commuters. Applications are accepted up to one week before the Exploration begins.

Summer Youth Explorations
Michigan Technological University
Youth Programs Office,
  Alumni House
1400 Townsend Drive
Houghton, MI 49931-1295
906-487-2219
http://youthprograms.mtu.edu/syp

Tourism Cares
Conferences/Field Experience/Volunteer Programs
This organization operates volunteer programs that "preserve, conserve, and protect tourism-related sites in America that need care and rejuvenation." Some of the places that Tourism Cares volunteers have recently helped restore include Louis Armstrong Park in New Orleans; Virginia City, Nevada (a town that played a major role during the silver rush in

the 1800s); and Mount Vernon (George Washington's estate and gardens).

Additionally, college students who are studying travel, tourism, or hospitality can participate in the Experience the Industry Student Program that allows them to shadow industry professionals (tour operators, tour suppliers, and destination management organizations) at the annual convention of the National Tour Association (the parent organization of Tourism Cares). This program allows students to learn more about career options in the field and make valuable connections with industry experts.

Tourism Cares
275 Turnpike Street, Suite 307
Canton, MA 02021-2357
781-821-5990
info@tourismcares.org
http://www.tourismcares.org

University Aviation Association
Conferences/Membership
The association offers membership for high school and college students. Membership benefits include access to job listings and scholarships, networking opportunities, a subscription to *Collegiate Aviation Review* (a quarterly newsletter), and discounts on association products. Members can also attend the association's conferences, including the Collegiate Aviation Summit. Visit its Web site for more information.

University Aviation Association
3410 Skyway Drive
Auburn, AL 36830-6444

334-844-2434
uaamail@uaa.aero
http://www.uaa.aero

U.S. Fish and Wildlife Service
Volunteer Programs
Volunteers of all ages are welcomed by the U.S. Fish & Wildlife Service. Volunteers may be tasked with leading tours, conducting wildlife population surveys, performing clerical tasks, photographing natural and cultural resources, helping with laboratory work, banding ducks at a National Wildlife Refuge, and restoring wildlife habitat. Contact the U.S. Fish & Wildlife Service for more information.

U.S. Fish and Wildlife Service
U.S. Department of the Interior
Division of Human Resources
4401 North Fairfax Drive, Room 634
Arlington, VA 22203-1610
800-344-9453
http://www.fws.gov/volunteers and
http://volunteer.gov/gov

Women in Aviation
Conferences/Field Experience/ Membership
This nonprofit organization offers membership options for aviation enthusiasts, youth, and college students. Membership is open to both females and males. Membership benefits include networking and mentoring opportunities, the chance to attend the organization's annual conference, a subscription to *Aviation for Women*, and access to scholarships and career and education resources.

Women in Aviation
Morningstar Airport
3647 State Route 503 South
West Alexandria, OH 45381-9354
937-839-4647
http://www.wai.org

Yosemite Institute
Camps/Field Experience/College Courses/Summer Study

The Yosemite Institute, established in 1971, works in cooperation with the National Park Service to offer several programs for youth. Overnight Wilderness Backpacking Trips are one- to three-night camping adventures that encourage young people to "appropriately challenge themselves while exploring the wilderness and practicing Leave No Trace ethics." Led by professional naturalist guides, you explore Yosemite's high peaks, deep canyons, alpine lakes, and other features rarely seen by other visitors. You learn about the area's abundant wildlife and unique cultural and natural history while hiking four to six miles per day at elevations of 6,000–10,000 feet. Only 12 participants are accepted for each Overnight Wilderness Backpacking Trip. The program fee includes meals and group overnight gear (tents, cooking pots, etc.). You must provide your own personal gear, however, including sleeping bag, water bottle, and utensils.

Teens who participate in the institute's two-week Field Research Course can earn college credit by creating their own ecology research project. You will learn "wilderness survival and backpacking skills; Sierra Nevada natural history and ecology; how to record field observations and identify patterns; how to generate answerable questions and hypotheses; how to collect data that will answer your question; and how to analyze and present your data to other scientists." To participate, you must be at least 16 years old and have completed at least one year of high school biology. The cost of the program is $1,900 for California residents and $2,400 for out-of-state participants. Participants will receive three college credits when they complete the program.

Additionally, young women between the ages of 15 and 18 can participate in the Armstrong Scholars Program, which "seeks to inspire young women to reach their highest potential and develop a stronger sense of self and community and a stronger connection to nature." The nine-day program costs $150 (the remaining costs are covered by a scholarship).

The Yosemite Institute also offers environmental workshops for teachers, and various programs throughout the year. Contact the institute for further information and for details on available scholarship funds.

Yosemite Institute
PO Box 487
Yosemite, CA 95389-0487
209-379-9511
yi@yni.org
http://www.yni.org/yi

Read a Book

When it comes to finding out about travel and tourism, don't overlook a book. (You're reading one now, after all.) What follows is a short, annotated list of books and periodicals related to travel and tourism. The books range from personal accounts of what it's like to work in the field to professional volumes on specific topics, such as how to run a bed and breakfast. Don't be afraid to check out the professional journals, either. The technical stuff may be way above your head right now, but if you take the time to become familiar with one or two, you're bound to pick up some of what is important to travel and hospitality professionals, not to mention begin to feel like a part of their world, which is what you're interested in, right?

We've tried to include recent materials as well as old favorites. Always check for the most recent editions, and, if you find an author you like, ask your librarian to help you find more. Keep reading good books!

❑ BOOKS

Alonzo, Roy S. *The Upstart Guide to Owning and Managing a Restaurant.* 2d ed. New York: Kaplan Business, 2007. Written by a professor of food service management, this book provides step-by-step strategies and action plans for understanding and managing restaurant staff, ensuring profitability, and creating ambiance.

Bock, Becky S. *Welcome Aboard!: Your Career as a Flight Attendant.* 3d ed. Newcastle, Wash.: Aviation Supplies & Academics Inc., 2005. An overview of what it's like to work as a flight attendant, describing the duties of the position as well as lifestyle adjustments those in the field must make. Special attention is given to preparing for job interviews.

Bow, Sandra. *Working on Cruise Ships.* 4th ed. Oxford, U.K.: Vacation Work Publications, 2005. Provides information on more than 150 career paths in the cruise industry, including captain, chef, bartender, engineer, nurse, photographer, fitness instructor, port lecturer, hairdresser, entertainer, and cruise director.

Burgett, Gordon. *Sell and Resell Your Magazine Articles.* Novato, Calif.: Communication Unlimited, 2002. Marketing methods for practicing writers with proven tips for selling journalistic articles, such as making one subject work for you in a variety of markets. Also explains how to develop an action plan for getting your work published.

Carter, Jimmy. *An Outdoor Journal: Adventures and Reflections.* Fayetteville, Ark.: University of Arkansas Press, 1994. The former president's funny and entertaining reflections on dangers in the woods, fly-fishing, learning to hunt, adventures in New Zealand and Kilimanjaro, and more.

Chmelynski, Carol Caprione. *Opportunities in Restaurant Careers.* New York: McGraw-Hill, 2004. Comprehensive book written by a veteran of food service. Provides a practical overview of the field, from education requirements to working conditions.

Conway, Richard, and Paul Tizzard. *Flying Without Fear: 101 Fear of Flying Questions Answered.* London, U.K.: Flying Without Fear Publishing, 2008. Provides answers to the questions most frequently asked by people who have a fear of flying. Great for aspiring pilots and travelers alike.

Cook, Andrew. *To Be an Airline Pilot.* Wiltshire, U.K.: The Crowood Press, 2007. The author, a licensed commercial airline pilot, takes you through all the steps leading up to assuming the controls of an airplane. The book includes a thorough description of the training process, coursework necessary for certification, and sample exam questions.

Craig, Patricia, ed. *The Oxford Book of Travel Stories.* New York: Oxford University Press USA, 2002. Brings together superb short fiction, from Charles Dickens and Evelyn Waugh to Anita Desai and Paul Theroux, about living and working abroad.

Eichenberger, Jerry A. *Your Pilot's License.* 7th ed. New York: McGraw-Hill Professional, 2003. Quintessential guide for aspiring and novice pilots. Outlines the ins and outs of flight education, from takeoff to landing.

Entrepreneur Press. *Start Your Own Travel Business and More.* Irvine, Calif.: Entrepreneur Press, 2007. Tips for starting your own home-based business in the competitive and changing travel/tourism industry, with an emphasis on "new travel trends" such as trips for senior citizens, cruises, "extreme" tours, and international travel.

Forsyth, Frederick, ed. *Great Flying Stories.* New York: W. W. Norton & Co., 1995. A collection of enchanting stories about flying by H. G. Wells, J. G. Ballard, Roald Dahl, Len Deighton, Sir Arthur Conan Doyle, and others.

Goeldner, Charles R., and J. R. Brent Ritchie. *Tourism: Principles, Practices, Philosophies.* 11th ed. Hoboken, N.J.: Wiley, 2008. A textbook for students in the tourism/hospitality industry that includes profiles of industry leaders, discusses the ways that economic and social factors affect tourism, and examines new developments in technology as well as the types of services that are available to customers.

Gorham, Ginger, and Susan Rice. *Travel Perspectives: A Guide to Becoming a Travel Professional.* 4th ed. Florence, Ky.: Delmar Cengage Learning, 2006. Aimed at aspiring and current travel professionals who need to stay on top of the many changes taking place in the industry, this reference includes information on the Internet as a business tool and government regulations relating to safety and security of travel.

Harvey, Mark. *The National Outdoor Leadership School's Wilderness Guide.* Rev. ed. New York: Fireside Press, 1999. Classic guide to wilderness camping and hiking techniques, with a focus on coping and conquering risks. Ideal for both guides and amateurs.

Henkin, Shepard. *Opportunities in Hotel & Motel Careers.* Rev. ed. New York: McGraw-Hill, 2006. A wealth of information for those interested in a career in the hotel/motel industry. Lists necessary skills, training requirements, typical duties, and average salaries for different specialties within the field, including desk clerks, chefs, guest services professionals, and those working in sales and management positions.

International Council on Hotel, Restaurant, and Institutional Education. *The Guide to College Programs in Hospitality, Tourism and Culinary Arts.* 9th ed. Richmond, Va.: International Council on Hotel, Restaurant, and Institutional Education, 2007. This book provides an overview of postsecondary programs in convention and meeting services, culinary arts, food service, hotel and lodging management, recreation services, restaurant management, and travel and tourism.

Kahn, Karen M. *Flight Guide for Success: Tips and Tactics for the Aspiring Airline Pilot.* Santa Barbara, Calif.: Pilot Career & Interview Counseling, 2004. Having worked as both an airline captain and an aviation career counselor, the author is uniquely positioned to provide honest, no-nonsense information on how to pursue a career as an airline pilot and what it will be like when you get there.

Krannich, Ron, and Caryl Krannich. *Jobs for Travel Lovers: Opportunities at Home and Abroad.* 5th ed. Manassas Park, Va.: Impact Publications, 2006. Great resource for the career-seeker who won't "settle down." Covers topics from teaching English abroad to transportation work.

Levine, Karen, and Alan Gelb. *A Survival Guide for Hotel and Motel Professionals.* Florence, Ky.: Delmar Cengage Learning, 2004. This book of helpful information on topics such as stress management and ways to increase efficiency can help those employed at hotels and motels from falling victim to the high rates of job turnover that are commonplace in the industry.

Mackay, Richard. *The Atlas of Endangered Species.* Berkeley, Calif.: University of California Press, 2008. This illustrated guide to the world's endangered species provides an overview of the major threats to biodiversity (loss of habitat, war, hunting, global warming) and the steps conservation scientists are taking to slow the destruction of wildlife. This book will be useful for tour guides, adventure-travel specialists, and other travel and tourism professionals.

Marks, Marsha. *Flying by the Seat of My Pants: Flight Attendant Adventures on a Wing and a Prayer.* Colorado

Springs, Colo.: WaterBrook Press, 2005. A series of humorous and sometimes moving stories from the author's 20 years as a flight attendant.

Masterson, Sky. *Pilots of the Line: On Being an Airline Pilot Before and Since 9-11-2001.* Bloomington, Ind.: iUniverse Inc., 2004. The author draws on his more than 15,000 hours of experience as an aviator in this collection of short stories that describe exactly what happens in the cockpit as well as what goes through the mind of an airline pilot.

McGavin, George C. *Endangered: Wildlife on the Brink of Extinction.* Richmond Hill, ON: Firefly Books, 2006. Provides an overview of endangered species, why they are at risk, and what we can do to save them. Includes more than 400 photographs.

Milne, Robert, and Marguerite Backhausen. *Opportunities in Travel Careers.* 2d ed. New York: McGraw Hill, 2003. Provides detailed information on a variety of careers with cruise lines, airlines, railroads, travel agencies, and other employers.

Mitchell, G. *How to Start a Tour Guiding Business.* Charleston, S.C.: BookSurge Publishing, 2005. An international expert in the travel/tourism field provides tips on entering and making a career in the industry, stressing the importance of marketing and promotion in the world's fastest growing profession.

Monaghan, Kelly. *Home-Based Travel Agent.* 5th ed. Branford, Conn.: The Intrepid Traveler, 2006. A practical guide to starting a home-based travel business, this book offers suggestions for finding one's niche or specialty and weighs the pluses and minuses of running a truly independent organization as opposed to working as a home-based agent of a larger company.

Ninemeier, Jack D., and Joe Perdue. *Hospitality Operations: Careers in the World's Greatest Industry.* Upper Saddle River, N.J.: Prentice Hall, 2004. If you are considering a career in the hospitality industry, this introductory textbook can help you decide if such a career is right for you and, if so, give you information on how to break into the field, achieve success within it, and develop a long-term career strategy.

O'Reilly, James, Larry Habegger, and Sean O'Reilly. *The Best Travel Writing 2008: True Stories from Around the World.* Palo Alto, Calif.: Travelers' Tales, 2008. A collection of award-winning stories about life on the road that will entertain and amuse armchair travelers and also inspire would-be authors to write down the stories of their own travels.

Painter, Amelia. *Opening and Operating a Bed & Breakfast in the 21st Century: Your Step-by-Step Guide to Inn Keeping Success with Professional Online Marketing Strategies.* 2d ed. Charleston, S.C.: BookSurge Publishing, 2007. If you are thinking about starting your own bed and breakfast business, or even if you have already started one, this helpful guide will give you valuable advice on how to best position yourself for success, with special attention given to marketing and increasing your business through the Internet.

Pasternak, Ceel. *Cool Careers for Girls in Travel and Hospitality.* Manassas Park, Va.: Impact Publications, 2003. This entry in the Cool Careers for Girls series contains profiles of 10 women who are employed in the travel/hospitality industry. These professionals tell the stories of how they got to where they are today and describe what to expect for girls who are interested in following a career path in the industry.

Pazden, Greg. *How to Become an Airline Pilot: Career Paths to the Airlines.* Scotts Valley, Calif.: CreateSpace, 2008. This book is a step-by-step guide to beginning and establishing a successful career as an airline pilot, including a series of "proven processes" for gaining a competitive edge in the field and overcoming obstacles of the profession, such as stress and solitude.

Potts, Rolf. *Marco Polo Didn't Go There: Stories and Revelations from One Decade as a Postmodern Travel Writer.* Palo Alto, Calif.: Travelers' Tales, 2008. An entertaining collection of stories from the author's journeys throughout Asia and Europe that, in addition to describing the sights, sounds, smells, and tastes of far-off places, offer insight into what it takes to make a living as a professional travel writer.

Quammen, David. *Wild Thoughts from Wild Places.* New York: Scribner, 1999. Fascinating, and often funny, essays about the natural world.

Rutherford, Denney G., and Michael J. O'Fallon, eds. *Hotel Management and Operations.* 4th ed. Hoboken, N.J.: Wiley, 2006. Comprehensive overview of the hospitality industry, covering housekeeping, front desk management and administration, telecommunications, food service, and other departments. It features interviews with more than 60 industry professionals.

Sandoval-Strausz, Andrew K. *Hotel: An American History.* New Haven, Conn.: Yale University Press, 2008. A comprehensive overview of the hotel and the hospitality industry from the early days of the United States through the present day, this book is a well-researched examination of the role of the hotel in American society and is thought-provoking reading for students in hospitality programs.

Shapiro, Michael. *A Sense of Place: Great Travel Writers Talk About Their Craft, Lives, and Inspiration.* Palo Alto, Calif.: Travelers' Tales, 2004. A fascinating resource for aspiring travel writers or even writers in general, this book contains interviews with 18 writers, including novelist Isabel Allende and guidebook authors Rick Steves and Arthur Frommer. Short excerpts from the writings of each subject are included.

Slaton, Hunter. *Vault Guide to the Top Hospitality & Tourism Industry Employers, 2009.* New York: Vault Inc., 2008. This entry in the Vault Guide series spotlights top employers in the hospitality/tourism industry, offering information on hiring practices, salary figures and the workplace culture of companies such as Marriott International, Trump Hotels, Continental Airlines, Hilton Hotels, and Royal Caribbean Cruises.

Strunk, William Jr. *The Elements of Style: 50th Anniversary Edition.* New York: Longman, 2008. A reissued, classic work on writing clear, precise, and concise prose. Of all the books on the craft of writing, this remains the essential guide by providing simple and direct rules that were first set forth by E. B. White's professor of English at Cornell University.

Ward, Kiki. *The Essential Guide to Becoming a Flight Attendant.* Colleyville, Tex.: Kiwi Productions, 2008. This book's detailed tips and role-playing strategies to help flight attendant applicants stand out in the job interview process are combined with a thorough description of what to expect after landing a flight attendant job.

White, Mary. *Running a Bed & Breakfast For Dummies.* Hoboken, N.J.: For Dummies, 2009. Every aspect of starting and running a profitable bed & breakfast business is covered in this resource for current and prospective entrepreneurs, including creating a business plan, handling the day-to-day responsibilities of ownership, and doing everything possible to ensure customer satisfaction and, hopefully, repeat business.

Whitelegg, Drew. *Working the Skies: The Fast-Paced, Disorienting World of the Flight Attendant.* New York: NYU Press, 2007. An in-depth study of the lives and careers of flight attendants, as pieced together from interviews with 60 professionals working in the field. This book gives an honest account of the day-to-day realities of the job and its perks as well as some of its drawbacks.

Wray, Cheryl. *Writing for Magazines: A Beginner's Guide.* New York: McGraw-Hill Humanities/Social Sciences/Languages, 2004. A complete look at what it takes to write magazine articles—how to assess your own personality profile, what types of articles to write, where to find ideas, how to manage your time, and everything you need to know about submitting manuscripts.

Zobel, Louise Purwin. *The Travel Writer's Handbook: How to Write—and Sell—Your Own Travel Experiences.* 6th ed. Evanston, Ill.: Surrey Books, 2006. The author, a college writing instructor and published travel writer, explains how to make money writing about your travel adventures. Topics include the importance of research, how to go about receiving assignments from publications, and using the Internet to further your writing goals.

❏ PERIODICALS

AdventureTravelNews. Published monthly by the Adventure Travel Trade Association (601 Union Street, 42nd Floor, Seattle, WA 98101-2341, 360-805-3131, info@adventuretravel. biz), this online magazine covers everything of interest to the adventure travel community, including profiles of resorts and tour providers that specialize in adventure travel and tips and news on responsible tourism. Visit http://www.adventuretravelnews.com to read sample articles.

ASTAnetwork. Published quarterly by the American Society of Travel Agents (1101 King Street, Alexandria, VA 22314-2944, http://www.asta.org), this member publication seeks to assist travel agencies with advice on management and professional development.

Aviation for Women. Published six times annually by Women in Aviation (Morningstar Airport, 3647 State Route 503 South, West Alexandria, OH 45381-9354, 937-839-4647, http://www.wai.org/magazine), this resource features news about the aviation industry, profiles of current and historical female aviation professionals, career advice, information on scholarships, and more.

Collegiate Aviation News. Published quarterly by the University Aviation Association (3410 Skyway Drive, Auburn, AL 36830-6444, 334-844-2434, uaamail@uaa.aero, http://www.uaa.aero), this newsletter features information on association activities, accreditation, scholarships, and industry and government developments.

Cornell Hospitality Quarterly. Published by Cornell University (800-818-7243, journals@sagepub.com), this journal provides an overview of industry developments for hotel and restaurant managers. Recent articles include "The Effects of Leadership Style on Hotel Employees' Commitment to Service Quality," "Scheduling Restaurant Workers to Minimize Labor Cost and

Meet Service Standards," and "Barometer of Hotel Room Revenue: A Regular Service of Smith Travel Research." Visit http://cqx.sagepub.com to read a sample issue.

Courier. Published monthly by the National Tour Association (546 East Main Street, Lexington, KY 40508-2342, 800-682-8886), this magazine for domestic and international tour operators contains in-depth information on destinations, travel business trends, product development tips, and case studies. Electronic editions are available at http://www.ntaonline.com/members/news-and-publications/courier.

Cruise Industry News Quarterly. Published by Cruise Industry News (441 Lexington Avenue, Suite 809, New York, NY 10017-3935, 212-986-1025, info@cruiseindustrynews.com), this resource provides information on a plethora of topics in the cruise industry, including "new ships, shipbuilding, naval architecture, design, repairs and refurbishments . . . marine and hotel operations, compliance, new technology, food and beverage, human resources, recruitment and training, safety and environmental protection, ports, destinations, and itinerary planning." Visit http://www.cruiseindustrynews.com/cruise-magazine.html to read sample articles.

Digital Traveler. Published monthly by The International Ecotourism Society (1301 Clifton Street, NW, Suite 200, Washington, DC 20009-7058, 202-506-5033, http://www.ecotourism.org), this is a free e-newsletter that provides updates and action alerts on ecotourism.

Earthwatch Journal. Member publication of the Earthwatch Institute (Three Clock Tower Place, Suite 100, Box 75, Maynard, MA 01754-2549, 800-776-0188, info@earthwatch.org, http://www.earthwatch.org/newsandevents/publications). Topics discussed range from acid rain to endangered species. Tour guides and adventure-travel specialists will find this information to be useful when planning trips and preparing presentations.

EcoCurrents. Published quarterly by The International Ecotourism Society (PO Box 96503, #34145, Washington, DC 20009-6503, 202-506-5033, http://www.ecotourism.org) for its members, each issue of this online magazine focuses on a specific theme within ecotourism and sustainable travel, such as new technologies, new destinations, and conservation education.

Executive Housekeeping Today. Published monthly by the National Executive Housekeepers Association (1001 Eastwind Drive, Suite 301, Westerville, OH 43081-3361, 800-200-6342, excel@ieha.org, http://www.ieha.org/magazine.php), this resource features profiles of industry leaders and helpful articles on industry topics that will be of interest to professional housekeepers. Recent articles include "A Greener Approach to Laundry Starts with the Little Things," "From Boom to Bust: Surviving Tight Economic Times," and "Big, Fat Green Weddings: Opportunities for the Hospitality Industry."

Flight Journal. Published bimonthly by Air Age Inc. (20 Westport Road, Wilton, CT 06897-4549, 800-829-9080, flightjournal@airage.com), this journal explores the ins and outs of various forms of aviation, from commercial to recreational. Visit http://www.flight-journal.com to read sample articles.

Flying. Published monthly by Hachette Filipacchi Media U.S. Inc. (1633 Broadway, 41st Floor, New York, NY 10019-6708, 212-767-4936). The world's most widely read aviation magazine, this is a very informative guide to anything from getting certified to the virtues of various flying careers. Sample articles are available at http://www.flyingmag.com.

GlobalWrites. This webzine is the official publication of the International Food, Wine and Travel Writers Association (1142 South Diamond Bar Boulevard, #177, Diamond Bar, CA 91765-2203). It features articles on tours, hotels, resorts, and restaurants by professional travel writers. The publication can be accessed by visiting http://www.global-writes.com.

Hosteur. Published biannually by the Council on Hotel, Restaurant and Institutional Education (2810 North Parham Road, Suite 230, Richmond, VA 23294-4422, 804-346-4800), this is the only international career and self-development magazine for future hospitality and tourism professionals. Used as a classroom and training resource by many educators. Recent articles include "Outer Space as a New Frontier for Hospitality and

Tourism," "The Accreditation Mania: What's the Difference in Standards," and "Leadership: The Foundation for Management." Visit http://www.chrie. org/i4a/pages/index.cfm?pageid=3391 to read a sample issue.

Hotel and Motel Management. Published 21 times a year (once a month in January, August, and December; two issues in every other month) by Questex Media Group Inc. (757 Third Avenue, 5th Floor, New York, NY 10017-2013, 212-895-8200), this magazine offers comprehensive coverage of the hotel industry, with articles and columns providing news analysis, sales and management tips, and discussions of legal and financing issues. Visit http://www.hotelworldnetwork. com/hotel_and_motel_management to view a sample issue.

Hotel Business. Published semi-monthly by ICD Publications Inc. (45 Research Way, Suite 106, East Setauket, NY 11733-6401, 631-246-9300, info@ hotelbusiness.com, http://www. hotelbusiness.com/hb/main.asp), this magazine contains information for hotel owners, general managers, and financial executives—the individuals involved in making decisions for all aspects of hotel ownership and operations. Print and electronic editions are available.

Hotelier. Published eight times a year by Kostuch Publications Limited (23 Lesmill Road, Suite 101, Toronto, ON Canada M3B 3P6, 416-447-0888), Canada's leading hotel publication features business developments, pro-files of industry movers and shakers, and in-depth features on issues facing the hotel business. Visit http://www. hoteliermagazine.com to read sample articles.

Innkeeping Quarterly. Published by the Professional Association of Innkeepers International (207 White Horse Pike, Haddon Heights, NJ 08035-1703, 800-468-7244), this publication features member profiles and covers topics such as finances, marketing, customer relations, real estate, food, operations, and more. Visit http://www.paii.org to read a sample issue.

Lodging. Published monthly by the American Hotel and Motel Association (385 Oxford Valley Road, Suite 420, Yardley, PA 19067-7723, 215-321-9662), this magazine covers lodging, hospitality, travel, and tourism from a professional perspective. Visit http://www.lodgingmagazine.com to sign up for a free subscription.

The National Culinary Review. Published monthly by the American Culinary Federation (180 Center Place Way, St. Augustine, FL 32095-8859, 800-624-9458), this resource features "chef-tested recipes, industry news, and culinary techniques, and is an educational resource for everyone interested in food preparation." Visit http://www.acfchefs.org to read a sample issue.

National Geographic. Published monthly by the National Geographic Society (PO Box 63002, Tampa, FL 33663-3002, 800-647-5463), this attractive and informative magazine provides a wealth of stories and features about environmental issues, plants and animals, and historical and cultural topics. Visit http://www.nationalgeographic.com to read sample articles.

National Geographic Adventure. Published 10 times annually by the National Geographic Society (PO Box 63002, Tampa, FL 33663-3002, 800-647-5463), this magazine for the general public features articles, reviews, and other information about adventure travel throughout the world. Recent articles include "50 Best American Adventures," "Gear: Buyer's Guide," and "World Class: Lodges + Escapes." Sample articles can be read by visiting http://adventure.nationalgeographic.com.

National Geographic Traveler. Published eight times annually by the National Geographic Society (PO Box 63002, Tampa, FL 33663-3002, 800-647-5463), this magazine provides fascinating articles about travel throughout the world, from Baltimore to Bali. Recent articles include "Tours of a Lifetime," "48 Hours Krakow," and "Traveling in Stride." Visit http://traveler.nationalgeographic.com to read sample articles.

National Parks. Published quarterly by the National Parks Conservation Association (1300 19th Street, NW, Suite 300, Washington, DC 20036-1628), this attractive publication for the general public includes information about national parks and reserves in the United States and the conservation of natural resources. To read sample articles, visit http://www.npca.org/magazine.

National Wildlife. Published six times annually by the National Wildlife Federation (11100 Wildlife Center Drive, Reston, VA 2019-5361, 800-822-9919), this is a popular magazine devoted to wildlife conservation issues. Visit http://www.nwf.org/nationalwildlife to read sample articles.

Outdoors Unlimited. Published monthly by the Outdoor Writers Association of America (121 Hickory Street, Suite 1, Missoula, MT 59801-1896, 406-728-7434), this online resource features industry news, writing tips, profiles of members, information on scholarships and contests, job listings, and association news. Visit http://owaa.org/publications.htm#OU to read a sample issue.

Outpost. Published bimonthly by Outpost Incorporated (425 Queen Street West, Suite 201, Toronto, ON M5V 2A5 Canada, 416-972-6635, info@outpostmagazine.com), this magazine provides interesting articles about "the good, the bad, and the ugly about travel, told in an honest, sometimes irreverent voice." Visit http://www.outpostmagazine.com to read sample articles.

Outside. Published monthly by Outside (400 Market Street, Santa Fe, NM 87501-7300, oumcustserv@cdsfulfillment.com), this is an essential magazine for adventurers and those in the adventure field; it explores tourism through profiles of outposts and expeditions, product and book reviews, and stories about great trips, thus illuminating the rewards and challenges faced by adventure travelers. Visit http://outside.away.com to read sample articles.

Plane & Pilot. Published 11 times annually by Werner Publishing Corporation (12121 Wilshire Boulevard, 12th Floor, Los Angeles, CA 90025-1123), this magazine addresses topics of interest to piston-engine pilots and others interested in private aviation and aircraft. Of great technical value in shedding light on aviation protocol and machinery. Visit http://www.planeandpilotmag.com to read sample articles.

Quill: The Magazine for Journalists. Published nine times annually by the Society of Professional Journalists (3909 North Meridian Street, Indianapolis, IN 46208-4011, 317-927-8000), this magazine offers articles on new developments in the field of journalism, with discussions of technology, professional standards, education, and law. Essential for practicing journalists. Visit http://www.spj.org/quillabout.asp to read sample articles.

TravelAgent. Published every two weeks by Questex Media Group Inc. (757 Third Avenue, 5th Floor, New York, NY 10017-2013, 212-895-8200), this magazine bills itself as "multimedia intelligence for travel professionals," and recent issues have focused on issues such as travel agent fraud, religious travel, and ecotourism, as well as spotlighted many different getaway locations. Sample issues can be viewed at http://www.travelagentcentral.com.

Travel Weekly. Published by Northstar Travel Media LLC (100 Lighting Way, Secaucus, NJ 07094-3626, 201-902-

2000), this publication offers comprehensive and impartial information, essential news, and knowledge put together by the world's largest supplier of travel information. Designed for agents and travelers alike. Recent articles include "Finding the Sweet Spots in the Travel Industry," "Spas: The Great Escape," and "Tourism Cares: Helping Preserve History at a Civil War Site." Sample articles can be accessed by visiting http://www.travelweekly.com.

U.S. Travel Insights. Published monthly by the U.S. Travel Association (USTA) (100 New York Avenue, NW, Suite 450, Washington, DC 20005-3934, 202-408-8422, feedback@tia.org), this newsletter is concerned mainly with news out of Washington that affects the travel industry. Subscription is free with USTA membership. Visit http://www.tia.org/researchpubs/publications.html to read sample issues.

U.S. Travel Outlook. Published monthly by the U.S. Travel Association (100 New York Avenue, NW, Suite 450, Washington, DC 20005-3934, 202-408-8422, feedback@tia.org), this newsletter by the USTA's senior vice president of research collects and presents facts, figures, and forecasts about the travel industry. Subscription is free with USTA membership. Visit http://www.tia.org/researchpubs/publications.html to read sample issues.

The Wayfarer. Published quarterly by the North American Travel Journalists' Association (3579 East Foothill Boulevard, Suite 744, Pasadena, CA 91107-3119, info@natja.org), this newsletter for writers, photographers, editors, and tourism professionals provides reports on travel journalism awards, travel-related humor, information on networking, and suggestions for trips. Sample issues can be viewed at http://www.natja. org/wayfarer.

Surf the Web

You must use the Internet to do research, to find out, to explore. The Internet is the closest you'll get to what's happening right now around the world. This chapter gets you started with an annotated list of Web sites related to travel and tourism. Try a few. Follow the links. Maybe even venture as far as asking questions in a chat room. The more you read about and interact with personnel in this field, the better prepared you'll be when you're old enough to participate as a professional.

One caveat: You probably already know that URLs change all the time. If a Web address listed below is out of date, try searching the site's name or other keywords. If it's still out there, you'll probably find it. If it's not, maybe you'll find something better!

❑ THE LIST

American Hotel and Lodging Association: Hospitality Schools and Programs
http://www.ahla.com/products_list_schools.asp

After activating a free registration, students considering entering a hospitality program can use this site's searchable database to learn more about schools and programs worldwide. Students already enrolled in hospitality programs can access a database of hotel groups, links to career Web sites, and articles on careers in the hospitality and tourism industries. There is also a collection of industry facts and figures, links to industry news and publications, and information on continuing education in hospitality and tourism.

AvJobs.com
http://www.avjobs.com

Although this site charges a fee for career advice and access to job listings, there is still plenty of free content here for those interested in a career in the aviation industry. There are overviews of virtually every job imaginable with airlines, at airports, and in aerospace; aviation career salary ranges; resume and interview tips; and job listings. The AVSchools section lists schools that specialize in aviation and allows students to ask questions directly to admissions counselors. You can also sign up for a free weekly newsletter that lists job opportunities at all levels in the airline and airport industry.

Career Voyages: Hospitality
http://www.careervoyages.gov/hospitality-main.cfm

Anyone considering working in the hospitality industry should check out this section of the Career Voyages site, a joint effort between the U.S. Departments of Labor and Education. It is broken down into two sections: Hotels and Lodging, and Restaurants and Food Service. Each contains an overview of the industry; a thorough list of education and training options; links to schools offering relevant training; a list of in-demand occupations for the industry; and other career information, including videos.

Cleared to Dream
http://www.clearedtodream.org

Sponsored by the Air Line Pilots Association, International, this site is a valuable resource created to show what it's like to be an airline pilot, what you need to do to achieve that goal, and what organizations are available to assist you in the process. There is a summary of education and certification requirements for different types of aircraft, a description of a day in the life of a pilot, a special section on women in aviation, salary information, and a comprehensive 24-page educational brochure that is a good starting point for anyone with the urge to fly. In addition to career guidance offered by the association, a section called Airline Aircraft gives examples of ways in which the association has lobbied successfully for changes to make air travel safer.

College Navigator
http://nces.ed.gov/collegenavigator

College Navigator is sponsored by the National Center for Education Statistics, an agency of the U.S. Department of Education. At the site, users can search for information on nearly 7,000 postsec-ondary institutions in the United States. Searches can be conducted by school name, state, programs/majors offered (including parks, recreation, and leisure studies), level of award, institution type, tuition, housing availability, campus settings, percentage of applicants who are admitted, test scores, availability of varsity athletic teams, availability of extended-learning opportunities, religious affiliation, and specialized mission. Additionally, users can export the results of their search into a spreadsheet, save the results of their session, and compare up to four colleges in one view. This is an excellent starting place to conduct research about colleges and universities.

CoolWorks.com
http://www.coolworks.com

Can you picture yourself saddling up burros at the Grand Canyon or working as a tour guide at Mount Rushmore this summer? CoolWorks quickly links you to a wealth of information about seasonal employment at dozens of national and state parks, preserves, monuments, and wilderness areas. There are also listings of jobs and volunteer opportunities at ski areas, private resorts, cruise ships, and summer camps. Most of the national and state jobs require that applicants be 18 years or older. Most national and state parks listed here have seasonal positions available. Specific job descriptions can also be accessed by searching a pull-down menu of U.S. states and regions or international locations. While only some jobs allow you to apply directly online, many have downloadable application forms.

Discover America
http://www.discoveramerica.com/ca

This is the official travel and tourism Web site of the United States. Its mission is to promote and facilitate increased travel to and within the United States, and while the site's sponsor, the U.S. Travel Association, is focused primarily on attracting international visitors, this is a strong resource for American travelers as well. You can read about destinations and view photos, videos, and maps of attractions in the United States and its territories, and there is further information available on many major cities and their attractions. Those who can't make up their minds about where to go can use the Activity Finder feature to help plan their next trip.

EnvironmentalCareer.com
http://environmental-jobs.com

EnvironmentalCareer.com provides environmental job listings in a variety of fields. Users can search by keyword, job category (such as ecotourism), employment type (full or part time, contract, internship, etc.), geographic area, and salary range.

Flight Attendant Facts.com
http://www.flightattendantsfacts.com

If you have ever been curious about what the life of a flight attendant is like, what it takes to become a flight attendant, and the career outlook for flight attendants, this site has all the information you need, and it comes straight from those already working in the industry. You can browse salary and benefit information, read what it's like to go through the necessary training, search for available jobs, read tips about how to stand out in a job interview, and learn how to prepare yourself for working with pilots once you are hired.

High School Journalism
http://www.highschooljournalism.org

All aspiring writers, including those who want to specialize in travel writing, should visit this site. Resources that will be of special interest to students include information on journalism schools, scholarships, and high school clubs; a glossary of terms; and links to more than 2,800 high school/teen news organizations. Links to training workshops and summer camps specializing in electronic journalism are available for teachers and students alike. There are also useful resources for school counselors.

Hospitality Net
http://www.hospitalitynet.org

Welcome to a site that aims to uncover all the nooks and crannies of the hospitality industry for you. The home page is straightforward and introduces you to the site's major sections. One of these sections—Industry News—dishes out breaking headlines in the industry, and provides a database where you can search past articles by category (such as Academic News, Franchise, and Technology News). Another section called HotelSchools offers a searchable database of postsecondary hospitality training programs throughout the world. There are also lists of useful books and columns by industry experts.

HotelJobs.com
http://www.hoteljobs.com

This site is devoted exclusively to hotel jobs and hospitality careers. If you are interested in the hospitality industry, HotelJobs.com can provide the exposure you need when seeking advancement in your career. Job seekers can post their resumes and search listings worldwide for employment at hotels, cruise ship lines, resorts, restaurants, and casinos. The site promises "the quickest, most effective, and most convenient ways for hotel and hospitality job seekers to post their resumes and search jobs."

How Stuff Works
http://www.howstuffworks.com

If you spend a lot of time wondering how stuff you use or see every day actually works, then this site should be on your short list of Web sites to explore. It covers how "stuff," as varied and timely as tsunamis, identity theft, and satellite radio, works. Complex concepts are carefully broken down and examined with photos and links to current and past news items about the subject. Topics of interest for those interested in travel and hospitality careers include How Becoming an Airline Pilot Works, How Airplanes Work, and How Adventure Travel Works, among others.

Let's Go Flying
http://www.aopa.org/letsgoflying

This site, which is sponsored by the Aircraft Owners and Pilots Association, is perfect for anyone "dreaming of flying." For students interested in aviation, there are plenty of articles, lists of job options, and salary data to guide them in considering the pursuit of a career in the field. If you are an interested reader of the site and are "ready to start," you can go on to learn about flight school programs and the time and cost involved in completing them. There is plenty of information to help you choose from the site's database of 3,500 flight schools. You can also sign up for a free monthly e-newsletter full of aviation-focused information that will help you get started with flying.

National Park Service: Nature and Science: Views of the National Parks
http://www.nature.nps.gov

Aspiring tour guide and adventure-travel specialists will find this to be an interesting Web site. Views of the National Parks is a multimedia program that "presents the natural, historical, and cultural wonders associated with national parks." Users can learn about parks in the system (such as the Badlands, Devil's Tower, and the Grand Canyon), listen to interviews about the parks and their features, and view photos of the parks.

NationJobs Network: Hotel, Restaurant & Travel
http://www.nationjob.com

This is the home page of the NationJobs Network, an online job search service that just happens to have oodles of positions posted for people looking for work in the hospitality and travel industries. If you're impatient with these large job databases, then go directly to http://

www.nationjob.com/hotel, where you'll find a list that's been whittled down to jobs in hospitality, travel, and entertainment. However, you might want to fine-tune your search even more by searching for a particular job or limiting the search to the part of the country where you live. If nothing catches your eye on your first visit to NationJobs, then take advantage of an excellent free service called P. J. (Personal Job) Scout. Just fill out a form and P. J. Scout will email you descriptions of jobs that just might be what you're seeking.

Outward Bound
http://www.outwardbound.org

The Outward Bound movement is dedicated to helping people learn, grow, and become more active in their communities after going through challenging experiences in the outdoors. The Outward Bound Web site lists upcoming wilderness expeditions and courses that are available for people from ages 12 and up, with a special section devoted to expeditions for at-risk youth. Some of the activities include backpacking, rock climbing, sailing, dog-sledding, and mountaineering. A section entitled OB//SESSION allows you to browse photos, videos, and stories of past expeditions and other activities posted by Outward Bound alumni and staff as well as some professional photographers and videographers.

Peterson's Summer Camps and Programs
http://www.petersons.com/summerop/code/ssector.asp

This Web site offers great information about academic and career-focused summer programs. Finding a camp that suits your interests is easy enough at this site; just search Peterson's database by activity (Academics, Arts, Sports, Wilderness/Outdoors, Special Interests), geographic region, category (Day Programs in the U.S., Residential Programs in the U.S., Travel in the U.S. and to Other Countries, Special Needs Accommodations), keyword, or alphabetically. Click on a specific program or camp for a quick overview description. In some instances you'll get a more in-depth description, along with photographs, applications, and online brochures.

TravelGreen
http://travelgreen.org

TravelGreen is the ultimate source for information on eco-friendly options and businesses within the travel community. The site tracks all the latest developments in green, or sustainable, travel, and has a list of case studies detailing the green policies of organizations from each sector of the industry, from corporate giants such as Harrah's and American Express to trade groups such as the American Restaurant Association and the American Hotel and Lodging Association.

Ask for Money

By the time most students get around to thinking about applying for scholarships, grants, and other financial aid, they have already extolled their personal, academic, and creative virtues to such lengths in essays and interviews for college applications that even their own grandmothers wouldn't recognize them. The thought of completing yet another application fills students with dread. And why bother? Won't the same five or six kids who have been competing for academic honors for years walk away with all the really good scholarships?

The truth is that most of the scholarships available to high school and college students are being offered because an organization wants to promote interest in a particular field, encourage more students to become qualified to enter it, and finally, to help those students afford an education. Certainly, having a great grade point average is a valuable asset. More often than not, however, grade point averages aren't even mentioned; the focus is on the area of interest and what a student has done to distinguish himself or herself in that area. In fact, sometimes the only requirement is that the scholarship applicant be studying in a particular area.

❑ GUIDELINES

When applying for scholarships there are a few simple guidelines that can help ease the process considerably.

Plan Ahead

The absolute worst thing you can do is wait until the last minute. For one thing, obtaining recommendations or other supporting data in time to meet an application deadline is incredibly difficult. For another, no one does his or her best thinking or writing under the gun. So get off to a good start by reviewing scholarship applications as early as possible—months, even a year, in advance. If the current scholarship information isn't available, ask for a copy of last year's version. Once you have the scholarship information or application in hand, give it a thorough read. Try to determine how your experience or situation best applies to the scholarship, or if it even fits at all. Don't waste your time applying for a scholarship in culinary arts if you hate making breakfast.

If possible, research the award or scholarship, including past recipients and, where applicable, the person in whose name the scholarship is offered. Often, scholarships are established to memorialize an individual who majored in hotel management or a related field, for example, but in other cases, the scholarship is to memorialize the *work* of an individual. In those cases, try to get a feel for the spirit of the person's work. If you have any similar interests, experiences, or abilities, don't hesitate to mention them.

Talk to others who received the scholarship, or to students currently studying in the same area or field of interest in which the scholarship is offered, and try to gain insight into possible applications or work related to that field. When you're working on the essay on why you want this scholarship, you'll have real answers—"I would benefit from receiving this scholarship because studying hotel management will help me become a better leader and improve my customer service skills."

Take your time writing the essays. Make sure that you are answering the question or questions on the application and not merely restating facts about yourself. Don't be afraid to get creative. Try to imagine what you would think of if you had to sift through hundreds of applications. What would you want to know about the candidate? What would convince you that someone was deserving of the scholarship? Work through several drafts and have someone whose advice you respect—a parent, teacher, or school counselor—review the essay for grammar and content.

Finally, if you know in advance which scholarships you want to apply for, there might still be time to stack the deck in your favor by getting an internship, volunteering, or working part time. Bottom line: The more you know about a scholarship, and the sooner you learn it, the better.

Follow Directions

Think of it this way: Many of the organizations that offer scholarships devote 99.9 percent of their time to something other than the scholarship for which you are applying. Don't make a nuisance of yourself by pestering them for information. Simply follow the directions as they are presented to you. If the scholarship application specifies that you should write for further information, then write for it—don't call.

Pay close attention to whether you're applying for a grant, a loan, an award, a prize, or a scholarship. Often these words are used interchangeably, but just as often they have different meanings. A loan is financial aid that must be paid back. A grant is a type of financial aid that does not require repayment. An award or prize is usually given for something you have done (built a park or helped distribute meals to the elderly); or something you have created (a musical composition, a design, an essay, a short film, a screenplay, or an invention). On the other hand, a scholarship is often a renewable sum of money that is given to a person to help defray the costs of college. Scholarships are given to candidates who meet the necessary criteria based on essays, eligibility, grades, or sometimes all three. They do not have to be paid back.

Supply all the necessary documents, information, and fees, and meet the deadlines. You won't win any scholarships by forgetting to include a recommendation from a teacher or failing to postmark the application by the deadline. Bottom line: Get it right the first time, on time.

Apply Early

Once you have the application in hand, don't dawdle. If you've requested it far enough in advance, there shouldn't be any reason for you not to turn it in well in advance of the deadline. You never know, if it comes down to two candidates, your timeliness just might be the deciding factor. Bottom line: Don't wait and don't hesitate.

Be Yourself

Don't make promises you can't keep. There are plenty of hefty scholarships available, but if they all require you to study something that you don't enjoy, you'll be miserable in college. And the side effects of switching majors after you've accepted a scholarship could be even worse. Bottom line: Be yourself.

Don't Limit Yourself

There are many sources for scholarships, beginning with your school counselor and ending with the Internet. All of the search engines have education categories. Start there and search by keywords, such as "financial aid," "scholarship," and "award." But don't be limited to the scholarships listed in these pages.

If you know of an organization related to or involved with the field of your choice, write a letter asking if they offer scholarships. If they don't offer scholarships, don't stop there. Write them another letter, or better yet, schedule a meeting with the executive director, education director, or someone in the public relations department and ask them if they would be willing to sponsor a scholarship for you. Of course, you'll need to prepare yourself well for such a meeting because you're selling a priceless commodity—yourself. Don't be shy, and be confident. Tell them

all about yourself, what you want to study and why, and let them know what you would be willing to do in exchange—volunteer at their favorite charity, write up reports on your progress in school, or work part time on school breaks and full time during the summer. Explain why you're a wise investment. Bottom line: The sky's the limit.

❏ ONE MORE THING

We have not listed financial aid that is available from individual colleges and universities. Why? There are two reasons. First, because there are thousands of schools that offer financial aid for students who are interested in studying hotel and restaurant management, aviation, or a related major, and we couldn't possibly fit them all in this book. Second, listing just a few schools wouldn't be helpful to the vast majority of students who do not plan to attend these institutions. This means it is up to you to check with the college that you want to attend for details about available financial aid. College financial aid officers will be happy to tell you what types of resources are available.

❏ THE LIST

Air Line Pilots Association, International
1625 Massachusetts Avenue, NW
Washington, DC 20036-2212
703-689-2270
http://www.alpa.org

The association offers a scholarship program for "sons or daughters of medically retired, long-term disabled, or deceased pilot members." Applicants must be high school seniors or college students and have a GPA of at least 3.0. Contact the association for more information.

Air Traffic Control Association

Attn: Scholarship Fund
1101 King Street, Suite 300
Alexandria, VA 22314-2963
703-299-2430
info@atca.org
http://www.atca.org/scholarship
  programfund.aspx

Students who are enrolled or accepted in an accredited postsecondary air traffic control or aviation-related program may apply for a variety of scholarships from the association. Applicants must demonstrate financial need and submit a certified transcript, two letters of recommendation, and an essay of no more than 400 words that addresses the following topic: How my education efforts will enhance my potential contribution to aviation. The application deadline is typically in early May. Visit the association's Web site to download an application.

American Academy of Chefs

c/o American Culinary Federation
180 Center Place Way
St. Augustine, FL 32095-8859
800-624-9458, ext. 102
academy@acfchefs.net
http://www.acfchefs.org/Content/
  NavigationMenu2/Schools/
  Scholarships/default.htm

The American Academy of Chefs, the honor society of the American Culinary Federation, offers scholarships to high school seniors and college students who are interested in pursuing careers in the culinary arts. High school applicants must have a GPA of at least 2.5 and be accepted to an accredited postsecondary institution, majoring in culinary or pastry arts, and have a career goal of becoming a chef or pastry chef. (Note: College-specific scholarships for students who are interested in studying culinary arts, baking and pastry arts, or hotel/restaurant management are also available.) Contact the academy for more information.

American Hotel and Lodging Association (AHLA)

1201 New York Avenue, NW,
  Suite 600
Washington, DC 20005-3931
202-289-3100
info@ahla.com
http://www.ahlef.org/content.
  aspx?id=19468

The association offers 10 academic scholarships for high school seniors and college students planning to or currently pursuing postsecondary study in hotel/restaurant management, culinary arts, travel/tourism, hotel administration, or related fields. High school seniors can apply for the Incoming Freshman Scholarship. Applicants must have a minimum GPA of 2.0 and be U.S. citizens or permanent U.S. residents. Preference will be given to graduates of the AHLA Educational Institute's Lodging Management

Program (http://www.lodgingmanagement.org). (Note: Some of the association's state partner organizations also offer scholarships.)

American Legion Auxiliary
8945 North Meridian Street
Indianapolis, IN 46260-5387
317-569-4500
alahq@legion-aux.org
http://www.legion-aux.org/
scholarships/index.aspx

Various state auxiliaries of the American Legion, as well as its national organization, offer scholarships to help students prepare for a variety of careers. Most require that candidates be associated with the organization in some way, whether as a child, spouse, etc., of a military veteran. Interested students should contact the auxiliary for further information.

Association on American Indian Affairs
Attn: Director of Scholarship
  Programs
966 Hungerford Drive, Suite 12-B
Rockville, MD 20850-1743
240-314-7155
lw.aaia@verizon.net
http://www.indian-affairs.org/
scholarships/aaia_scholarships.htm

Undergraduate and graduate Native American students who are pursuing a wide variety of college majors can apply for several different scholarships of $1,500. All applicants must provide proof of Native American heritage. Visit the association's Web site for more information.

CollegeBoard: Scholarship Search
http://apps.collegeboard.com/
cbsearch_ss/welcome.jsp

This testing service (PSAT, SAT, etc.) also offers a scholarship search engine at its Web site. It features scholarships worth a total of nearly $3 billion. You can search by specific major (such as hospitality) and a variety of other criteria.

CollegeNET: MACH 25-Breaking the Tuition Barrier
http://www.collegenet.com/mach25/
app

CollegeNET features 600,000 scholarships worth more than $1.6 billion. You can search by keyword (such as "travel" or "hotel") or by creating a personality profile of your interests.

FastWeb
http://fastweb.monster.com

FastWeb is one of the best-known scholarship search engines around. It features 1.3 million scholarships worth a total of more than $3 billion. To use this resource, you will need to register (free).

Foundation for the Carolinas
Office of Scholarships
217 South Tryon Street
Charlotte, NC 28202-3201
704-973-4537
tcapers@fftc.org
http://www.fftc.org

The foundation administers more than 105 scholarship programs that offer awards to high school seniors and undergraduate and graduate students who plan to or who are currently pursuing study in a variety of disciplines. Visit its Web site for a list of awards.

GuaranteedScholarships.com
http://www.guaranteed-
scholarships.com

This Web site offers lists (by college) of scholarships, grants, and financial aid that "require no interview, essay, portfolio, audition, competition, or other secondary requirement."

Hawaii Community Foundation
1164 Bishop Street, Suite 800
Honolulu, HI 96813-2817
888-731-3863
info@hcf-hawaii.org
http://www.hawaiicommunity
foundation.org/scholar/scholar.php

The foundation offers a variety of scholarships for high school seniors and college students planning to or currently studying a variety of majors. Applicants must be residents of Hawaii, demonstrate financial need, and attend a two- or four-year college. Visit the foundation's Web site for more information and to apply online.

Hispanic College Fund (HCF)
1301 K Street, NW,
   Suite 450-A West
Washington, DC 20005-3317
800-644-4223
hcf-info@hispanicfund.org
http://www.hispanicfund.org

The Hispanic College Fund, in collaboration with several major corporations, offers many scholarships for high school seniors and college students planning to or currently attending college. Applicants must be Hispanic, live in the United States or Puerto Rico, and have a GPA of at least 3.0 on a 4.0 scale. Contact the HCF for more information.

Illinois Career Resource Network
http://www.ilworkinfo.com/icrn.htm

Created by the Illinois Department of Employment Security, this useful site offers a scholarship search engine, as well as detailed information on careers (including travel-related jobs). You can search for travel- and tourism-oriented scholarships based on major (such as aviation operations and services, business, hospitality management, travel services management, etc.) and other criteria. This site is available to everyone, not just Illinois residents; you can get a password by simply visiting the site. The Illinois Career Resource Network is just one example of the type of sites created by state departments of employment security (or departments of labor) to assist students with financial- and career-related issues. After checking out this site, visit your state's department of labor Web site to see what it offers.

Imagine America Foundation
1101 Connecticut Avenue, NW,
   Suite 901
Washington, DC 20036-4303
202-336-6800
http://www.imagine-america.org/
   scholarship/a-about-scholarship.asp

The Imagine America Foundation (formerly the Career College Foundation) is a nonprofit organization that helps students pay for college. It offers three $1,000 scholarships each year to high school students or recent graduates. Applicants must have a GPA of at least 2.5 on a 4.0 scale, have financial need, and perform voluntary community service during their senior year. Scholarships can be used at more than 500 career colleges in the United States. These colleges offer a variety of topics and fields of study, including accounting, air traffic controller, business administration, commercial pilot, entrepreneurship, food preparation, food services technology, franchising, hospitality administration, hospitality and recreation marketing operations, hotel/motel management, human resources management, international business, journalism, marketing, restaurant management, sales, security and loss prevention services, tourism and travel services management, and tourism promotion operations. Visit the foundation's Web site for more information.

James Beard Foundation Scholarship Program
Six West 18th Street, 10th Floor
New York, NY 10011-4608
212-627-1128
dhbrown@jamesbeard.org
http://www.jamesbeard.org

High school seniors and undergraduates who are interested in pursuing a career in the culinary arts may apply for a variety of scholarships, which are awarded based on financial need, academic achievement, and other criteria. Applications are typically due in mid-May. Visit the founda-

tion's Web site for more information and to download an application.

Leroy Homer Foundation
800-388-1647
http://www.leroywhomerjr.org/scholarships

The foundation offers a scholarship to help young people earn their private pilot license. Applicants must be U.S. citizens or permanent residents and be between the age of 16 and 23 years old by the application deadline. Applicants must not be participating in a training program that is offered by a college or university program. Visit the foundation's Web site to download an application.

Marine Corps Scholarship Foundation
PO Box 3008
Princeton, NJ 08543-3008
800-292-7777
mcsfnj@mcsf.org
http://www.marine-scholars.org

The foundation provides children of marines and former marines with scholarships of up to $4,500 for postsecondary study. To be eligible, you must be a high school graduate or registered as an undergraduate student at an accredited college or vocational/technical institute. Additionally, your total family gross income may not exceed $80,000. Contact the foundation for further details.

National Restaurant Association Educational Foundation
Attn: Scholarship Program
175 West Jackson Boulevard,
    Suite 1500

Chicago, IL 60604-2702
800-765-2122
info@restaurant.org
http://www.nraef.org/scholar_ment/
   sm_high_school.asp

High school seniors, GED graduates, and first-time college students (regardless of age) who are interested in studying culinary arts, restaurant management, or another food-service-related major and pursuing a career in the restaurant and food service industry are eligible for $2,500 scholarships. Visit the foundation's Web site for details on application requirements and to download an application.

Sallie Mae

http://www.collegeanswer.com/
   paying/scholarship_search/pay_
   scholarship_search.jsp

Sallie Mae offers a scholarship database of more than 3 million awards worth a total of more than $16 billion. You must register (free) to use the database.

Scholarship America

One Scholarship Way
PO Box 297
Saint Peter, MN 56082-0297
800-537-4180
http://www.scholarshipamerica.org

This organization works through its local Dollars for Scholars chapters throughout the United States. In 2008 it awarded more than $219 million in scholarships to students. Visit Scholarship America's Web site for more information.

Scholarships.com

http://www.scholarships.com

Scholarships.com offers a free college scholarship and grant search engine (although you must register to use it) and financial aid information. Its database of awards features 2.7 million listings worth a total of more than $19 billion in aid.

Tourism Cares

Attn: Scholarship Program Director
275 Turnpike Street, Suite 307
Canton, MA 02021-2357
781-821-5990
info@tourismcares.org
http://www.tourismcares.org

This organization's motto is "Restoring the past. Preserving the future." It operates volunteer programs that restore "culturally significant, tourism-related sites," and awards more than $100,000 each year to college students. The organization manages more than 50 scholarships for the National Tour Association and the American Society of Travel Agents.

United Negro College Fund (UNCF)

8260 Willow Oaks Corporate Drive
PO Box 10444
Fairfax, VA 22031-8044
800-331-2244
http://www.uncf.org/forstudents/
   scholarship.asp

Visitors to the UNCF Web site can search for information on thousands of scholarships and grants, many of which are administered by the UNCF. The site's search engine allows you to search by major (such as accounting, advertising, business, communications, English, food service, general, hotel management, human resources, journalism, management, and restaurant management), state, scholarship title, grade level, and achievement score. High school seniors and undergraduate and graduate students are eligible.

U.S. Department of Education
Federal Student Aid
800-433-3243
http://www.federalstudentaid.ed.gov
http://studentaid.ed.gov/students/
   publications/student_guide/index.
   html

The U.S. government provides a wealth of financial aid in the form of grants, loans, and work-study programs. Each year it publishes *Funding Education Beyond High School*, a guide to these funds. Visit the Web sites above for detailed information on federal financial aid.

Women in Aviation, International
Morningstar Airport
3647 State Route 503 South
West Alexandria, OH 45381-9354
937-839-4647
dwallace@wai.org
http://www.wai.org/education/
   scholarships.cfm

Women in Aviation, International offers more than 30 scholarships for high school seniors and college students who are interested in aviation, many of which are sponsored by corporations and other organizations. Contact the organization for complete details.

Look to the Pros

The following professional organizations offer a variety of materials, from career information, to lists of accredited schools, to salary surveys. Many also publish journals and newsletters that you should become familiar with. Some also have annual conferences that you might be able to attend. (While you may not be able to attend a conference as a participant, it may be possible to "cover" one for your school or even your local paper, especially if your school has a related club.)

When contacting professional organizations, keep in mind that they all exist primarily to serve their members, be it through continuing education, professional licensure, political lobbying, or just "keeping up with the profession." While many are strongly interested in promoting their profession and passing information about it to the general public, these busy professional organizations do not exist solely to provide you with information. Whether you call or write, be courteous, brief, and to the point. Know what you need and ask for it. If the organization has a Web site, check it out first; what you're looking for may be available to download, or you may find a list of prices or instructions, such as sending a self-addressed stamped envelope with your request. Finally, be aware that organizations, like people, move. To save time when writing, first confirm the address, preferably with a quick phone call to the organization itself: "Hello, I'm calling to confirm your address. . . ."

❏ THE SOURCES

Adventure Travel Trade Association
601 Union Street, 42nd Floor
Seattle, WA 98101-2341
360-805-3131
info@adventuretravel.biz
http://www.adventuretravel.biz

This organization represents the professional interests of companies in the adventure- travel industry. Visit its Web site for information on sustainable adventure travel, job listings, and to read sample articles from *AdventureTravelNews*.

Air Line Pilots Association, International
1625 Massachusetts Avenue, NW
Washington, DC 20036-2212
703-689-2270
http://www.alpa.org

Visit the association's Web site for a list of flight schools and information on careers, flight training, and financial aid for children of members.

Air Traffic Control Association
1101 King Street, Suite 300

Alexandria, VA 22314-2963
703-299-2430
info@atca.org
http://www.atca.org

The association represents air traffic controllers in the United States. Visit its Web site for information on scholarships for high school and college students.

Air Transport Association of America

1301 Pennsylvania Avenue, NW, Suite 1100
Washington, DC 20004-1738
202-626-4000
ata@airlines.org
http://www.airlines.org

This organization represents the leading airlines in the United States. Visit its Web site for job listings and to read *The Airline Handbook* and an aviation glossary.

American Culinary Federation

180 Center Place Way
St. Augustine, FL 32095-8859
800-624-9458
academy@acfchefs.net
http://www.acfchefs.org

This is a professional organization for chefs and cooks. Visit its Web site for information on careers, training programs, certification, and scholarships and membership for high school and college students.

American Hotel and Lodging Association

1201 New York Avenue, NW, Suite 600
Washington, DC 20005-3931

202-289-3100
eiinfo@ahla.com
http://www.ahla.com

Visit the association's Web site for job listings, a database of educational programs, and information on hotel careers, college student membership, job shadowing, scholarships, and its Lodging Management Program, an advanced curriculum for high school students.

American Hotel and Lodging Educational Institute

800 North Magnolia Avenue, Suite 300
Orlando, FL 32803-3261
800-752-4567
http://www.ei-ahla.org

Visit the institute's Web site for information on scholarships, certification, and programs for high school students.

American Society of Travel Agents

1101 King Street, Suite 200
Alexandria, VA 22314-2963
703-739-2782
askasta@asta.org
http://www.astanet.com

The society offers a salary tool, a list of educational programs, job listings, and information on scholarships and membership for postsecondary students at its Web site.

Association of Flight Attendants-CWA

501 Third Street, NW
Washington, DC 20001-2760
202-434-1300

info@afacwa.org
http://www.afanet.org

This union represents more than 55,000 flight attendants at 20 airlines. It offers scholarships to children of its members.

Association of Professional Flight Attendants
1004 West Euless Boulevard
Euless, TX 76040-5009
800-395-2732
http://www.apfa.org

This union represents more than 18,000 flight attendants who are employed by American Airlines.

Cruise Lines International Association
910 SE 17th Street, Suite 400
Fort Lauderdale, FL 33316-2968
754-224-2200
info@cruising.org
http://www.cruising.org

This is the official trade organization of the cruise industry. Visit its Web site for an overview of the field.

Dow Jones Newspaper Fund (DJNF)
PO Box 300
Princeton, NJ 08543-0300
609-452-2820
djnf@dowjones.com
https://www.newspaperfund.org

The DJNF provides general information on journalism careers, colleges and universities with journalism programs, scholarships, and job listings at its Web site. One of its most popular publications is *Journalist's Road to Success: A Career Guide*, which provides an overview of newspaper jobs (including those in online journalism), describes how to prepare for a journalism career, and provides information on internships and professional organizations.

Earthwatch Institute
Three Clock Tower Place, Suite 100
PO Box 75
Maynard, MA 01754-2549
800-776-0188
info@earthwatch.org
http://www.earthwatch.org

This organization offers international environmental expeditions that educate people about biodiversity, sustainability, habitat loss, coral reef health, indigenous cultures, climate change, and other environmental issues. Students can become members of the institute and participate in summer activities.

Federal Aviation Administration
800 Independence Avenue, SW
Washington, DC 20591-0001
866-835-5322
http://www.faa.gov

This federal agency regulates civil air transportation in the United States. Visit its Web site for information on pilot training schools, links to aviation-related scholarships, career information, and employment and internship opportunities for teens.

Hospitality Financial and Technology Professionals
11709 Boulder Lane, Suite 110
Austin, TX 78726-1832

800-646-4387
http://www.hftp.org

This organization represents nearly 5,000 finance and technology professionals in the hospitality industry. Visit its Web site for information on certification, publications, and membership and scholarships for college students.

Hospitality Sales and Marketing Association International

1760 Old Meadow Road,
  Suite 500
McLean, VA 22102-4306
703-506-3280
info@hsmai.org
http://www.hsmai.org

Visit the association's Web site for a list of hospitality degree programs, job listings, and information on scholarships and membership for college students.

International Council on Hotel, Restaurant, and Institutional Education

2810 North Parham Road, Suite 230
Richmond, VA 23294-4422
804-346-4800
info@chrie.org
http://chrie.org

Visit the council's Web site for information on purchasing a directory of educational programs, tips on finding a job, and to read a sample issue of *Hosteur* magazine.

The International Ecotourism Society

PO Box 96503, #34145
Washington, DC 20009-6503
202-506-5033

info@ecotourism.org
http://www.ecotourism.org

This nonprofit organization is "committed to helping organizations, communities, and individuals promote and practice the principles of ecotourism." Its members include academics, governments, architects, consultants, conservation professionals and organizations, lodge owners and managers, tour operators, general development experts, and ecotourists. Visit its Web site for information on the ecotourism industry, education and training, volunteer opportunities, membership, and *EcoCurrents*.

International Executive Housekeepers Association

1001 Eastwind Drive, Suite 301
Westerville, OH 43081-3361
800-200-6342
excel@ieha.org
http://www.ieha.org

Contact the association for information on careers, education, certification, *Executive Housekeeping Today*, and scholarships for college students.

International Food, Wine and Travel Writers Association

1142 South Diamond Bar Boulevard,
  #177
Diamond Bar, CA 91765-2203
877-439-8929
admin@ifwtwa.org
http://www.ifwtwa.org

Visit the association's Web site to read blogs, articles about the field, and member profiles. Membership for college students is also available.

Les Clefs d'Or USA

68 Laurie Avenue
Boston, MA 02132-5539
617-469-5397
info@lcdusa.org
http://lcdusa.org

Contact this organization for information on concierge careers and opportunities.

National Association of Career Travel Agents

1101 King Street, Suite 200
Alexandria, VA 22314-2963
877-22-NACTA
nacta@nacta.com
http://www.nacta.com

This organization represents independent contractors, cruise- and tour-oriented agents, outside-sales agents, and group-oriented travel professionals. It is affiliated with the American Society of Travel Agents.

National Business Travel Association

110 North Royal Street, 4th Floor
Alexandria, VA 22314-3234
703-684-0836
info@nbta.org
http://www.nbta.org

This organization represents travel managers and executives who seek to "balance employee needs with corporate goals, financial and otherwise." Visit its Web site for job listings, information on certification, and statistics on business travel.

National Concierge Association

612-253-5110

info@nationalconciergeassociation.com
http://www.nationalconcierge association.com

The association represents concierges who work in "corporate, hotel, retail, entertainment, academic, civic, medical, residential, and privately owned sectors of the hospitality industry." Visit its Web site for information on certification.

National Restaurant Association

1200 17th Street, NW
Washington, DC 20036-3006
202-331-5900
http://www.restaurant.org

This organization represents the professional interests of the U.S. restaurant industry. Visit its Web site for information on careers, education, certification, industry trends, and scholarships for high school and college students.

National Ski Areas Association

133 South Van Gordon Street, Suite 300
Lakewood, CO 80228-1706
303-987-1111
nsaa@nsaa.org
http://www.nsaa.org

This trade organization "represents 329 alpine resorts that account for more than 90 percent of the skier/snowboarder visits" in the United States. Visit its Web site for industry statistics.

National Tour Association

546 East Main Street
Lexington, KY 40508-2342
800-682-8886

questions@ntastaff.com
http://www.ntaonline.com

This organization represents professionals who are employed in the packaged-travel industry. Visit its Web site for information on certification, a glossary of terms, and to read *Courier* magazine.

National Writers Union (NWU)

113 University Place, 6th Floor
New York, NY 10003-4527
212-254-0279
nwu@nwu.org
http://www.nwu.org

Visit the NWU Web site for information about working as a writer and union membership.

Newspaper Association of America

4401 Wilson Boulevard, Suite 900
Arlington, VA 22203-1867
571-366-1000
http://www.naa.org

The association "represents the $47 billion newspaper industry and more than 2,000 newspapers in the U.S. and Canada." Visit its Web site for information on careers and industry trends.

North American Travel Journalists Association

150 South Arroyo Parkway,
    2nd Floor
Pasadena, CA 91105-4150
626-376-9754
http://www.natja.org

This is a professional membership organization for "working travel journalists."

Visit its Web site for information on publications such as *The Wayfarer*.

Outdoor Industry Association

4909 Pearl East Circle, Suite 200
Boulder, CO 80301-2499
303-444-3353
info@outdoorindustry.org
http://www.outdoorindustry.org

This nonprofit organization represents companies in the active outdoor recreation business. Visit its Web site for press releases, adventure travel news, and job listings.

Outdoor Writers Association of America

121 Hickory Street, Suite 1
Missoula, MT 59801-1896
406-728-7434
http://owaa.org/index.php

Members of the association include writers, book authors, editors, broadcasters, film and video producers, artists, photographers, and lecturers. Visit its Web site for information on publications and membership and scholarships for college students. The association also offers a writing competition for students in grades six through 12.

Professional Association of Innkeepers International

207 White Horse Pike
Haddon Heights, NJ 08035-1703
800-468-7244
http://www.paii.org

This organization represents the professional interests of bed and breakfast and country inn owners. Visit its Web site for career information, publications, and statistics about the industry. The association also offers an aspiring innkeeper membership option (although its cost—$199—suggests that it is geared more toward young professionals or career changers).

Seafarers International Union

5201 Auth Way
Camp Springs, MD 20746-4211
301-899-0675
http://www.seafarers.org

The union represents workers who are employed in deck, engine, and steward positions on passenger ships, gaming vessels, commercial containerships and tankers, military support ships, tugboats and barges, and other vessels.

Society of American Travel Writers

7044 South 13th Street
Oak Creek, WI 53154-1429
414-908-4949
satw@satw.org
http://www.satw.org

This is a professional organization for writers, electronic media and journalists, film lecturers, photographers, editors, broadcast/video/film producers, and public-relations representatives. Visit its Web site for tips on travel writing and photography.

Society of Government Travel Professionals

4938 Hampden Lane, #332
Bethesda, MD 20814-2914

202-363-7487
govtvlmkt@aol.com
http://www.sgtp.org

Contact the society for information on travel careers in the U.S. government.

Society of Professional Journalists

Eugene S. Pulliam National Journalism Center
3909 North Meridian Street
Indianapolis, IN 46208-4011
317-927-8000
http://www.spj.org

The society has chapters for college students all over the United States. Among its many services to students, it offers information on careers and internships.

Travel and Tourism Research Association

3048 West Clarkston Road
Lake Orion, MI 48362-2052
248-708-8872
admin@ttra.com
http://www.ttra.com

This nonprofit is "committed to improving the quality, value, effectiveness, and use of travel and tourism research and marketing information." Visit its Web site for statistics and other research on the travel and tourism industry.

The Travel Institute

148 Linden Street, Suite 305
Wellesley, MA 02482-7916
800-542-4282
info@thetravelinstitute.com
http://www.thetravelinstitute.com

Visit the institute's Web site for information on certification and continuing education for travel industry professionals, membership for college students, job listings, and career advice.

United States Tour Operators Association
275 Madison Avenue, Suite 2014
New York, NY 10016-1101
212-599-6599
information@ustoa.com
http://www.ustoa.com

Visit the association's Web site for information on the tour industry, a travel glossary, and answers to common questions about the industry.

University Aviation Association
3410 Skyway Drive
Auburn, AL 36830-6444
334-844-2434
uaamail@uaa.aero
http://www.uaa.aero

The association offers membership for high school and college students, scholarships for college students, information on member colleges, a design competition for college students, and other resources. Visit its Web site for more information.

U.S. Travel Association
1100 New York Avenue, NW,
   Suite 450
Washington, DC 20005-3934
202-408-8422
http://www.tia.org

This nonprofit trade organization "represents and speaks for the common interests of the $740 billion U.S. travel industry." Visit its Web site for travel statistics, job listings, and information on scholarships for college students.

Women in Aviation
Morningstar Airport
3647 State Route 503 South
West Alexandria, OH 45381-9354
937-839-4647
http://www.wai.org

This nonprofit organization is "dedicated to providing networking, education, mentoring, and scholarship opportunities for women (and men) who are striving for challenging and fulfilling careers in the aviation and aerospace industries." Visit its Web site for information on scholarships for high school and college students, membership for aviation enthusiasts and college students, publications, and other resources.

World Tourism Organization (WTO)
Capitán Haya 42
28020 Madrid, Spain
omt@unwto.org
http://www.unwto.org

Visit the WTO Web site for statistical information and industry news.

Index

Entries and page numbers in **bold** indicate major treatment of a topic.